THE PRENTICE-HALL

# CONCISE BOOK OF ASTRONOMY

THE PRENTICE-HALL CONCISE BOOK OF

# ASTRONOMY

by Jacqueline & Simon Mitton, University of Cambridge

Prentice-Hall, Inc., Englewood Cliffs, New Jersey

Concise Book of Astronomy



First American edition published 1979 by Prentice-Hall, Inc.

Created, designed and produced by
Trewin Copplestone Publishing Ltd, London.



Phototypeset by SX Composing Ltd, Rayleigh, Essex

Printed in Italy by New Interlitho SPA Milan.

10 9 8 7 6 5 4 3 2 1

Library of Congress Cataloging in Publication Data

Mitton, Jacqueline.
Concise book of astronomy.

SUMMARY: Presents an introduction to astronomy and the universe. Includes suggestions for simple observations and projects such as observing bright stars and making a shadow clock.
1. Astronomy—Juvenile literature. [1. Astronomy. 2. Universe] I. Mitton, Simon, 1946– joint author. II. Title.

QB46.M6 520 78–9695
ISBN 0–13–166967–2

# Contents

# The World of Astronomy

On a clear night the dark sky glitters with the sparkling stars and the silvery haze of the Milky Way. All through history, books and poems, paintings and people's religious beliefs have been inspired by the Sun, Moon and stars. What is a star? Where are they? How large is the Universe? Where did the Sun and Earth come from? These are some of the questions that have fascinated people for thousands of years. Astronomy is the science that tries to answer these questions.

We only know about the vastness of our Universe because we can see it! If we had the misfortune to live on a planet where the skies are perpetually cloudy, such as Venus, we would never see even the Sun, let alone the other planets and stars. Our Earth is 150 million kilometers from the Sun. Further out there are the giant planets Jupiter and Saturn, nearer in the dwarf worlds of Venus and Mercury. The Sun and its retinue formed from a cloud of gas about five thousand million years ago. In a further five thousand million years the Sun will swell up and engulf its family of planets.

Our Sun happens to be a very ordinary star. By studying the Sun, astronomers have learned how stars work. Since the next nearest star is over forty million million kilometers distant, the Sun gives us a unique opportunity to get a closeup view of a star.

The Sun is just one tiny speck in a vast starry family named the Milky Way, or the Galaxy. It contains about one hundred thousand million stars of various types, sprawled across a hundred thousand light years of space. (A light year is ten million million kilometers, the distance light travels in a year.) The Galaxy is a giant whirlpool of stars that has existed for twelve thousand million years. It was already very ancient before our Sun even came into being.

Giant leaps take us beyond the limits of the Milky Way to neighboring galaxies. Now every step we take must be measured in millions of light years. Two million light years bring us to another huge galaxy, the Andromeda nebula; ten million bring two dozen large galaxies into view. At fifty million light years we reach a whole cluster of galaxies milling around as one family. Beyond, countless billions of other galaxies stretch out of sight to the far boundary of the cosmos.

From our tiny planet, orbiting an ordinary star, we view this unimaginably vast Universe of stars and galaxies. The astronomer's task is to understand the nature of the Universe and the things in it. We have written this book as a basic guide to astronomy. It is intended for everyone who wants to know what astronomers do and what we have already discovered about the planets, stars and galaxies.

In this book we have used the metric system of units, as used by all scientists in their everyday work. For those unfamiliar with this system a few conversions may help: one *kilometer* is a little less than two-thirds of a *mile* and slightly over 1100 *yards*. In kilometers the distance from the Earth to the Sun is 150 million, whereas in miles it is 93 million. The metric *tonne* is almost identical to the imperial *ton*. For temperature measurement we use the Centigrade unit: 10,000°F is about 5300°C; the surface temperature of the Sun is 11,200°F or 6000°C. In the case of telescope sizes we have given the imperial size in brackets since many of the older instruments are still generally referred to by a name that includes an imperial unit. Thus the 5-meter reflector at the Mount Palomar Observatory is often called by its original name: "The 200-inch" telescope.

# Day and Night

Everyday the Sun rises over the eastern horizon. It climbs higher in the sky, following a curved path. In the evening it sets in the west and nightfall soon follows. Man, animals and plants are influenced by the daily cycle. Some animals rest by day and seek their food at night. Others, like people, do the opposite. The flowers of many plants open in daylight and close at night. But what causes "day" and "night?"

In ancient times, people thought that the Sun itself actually travelled across the heavens. The Egyptians, for example, had their sun-god, Ra, who daily rode a fiery chariot across the skies. Today we realize that it is not the Sun that moves, but the Earth. Our planet is like a spinning ball, turning from west to east. Once-per-day rotation makes the Sun appear to move across our sky. When we are on the side of the Earth facing the Sun we have day. At the same time it is night on the half of the Earth facing away from the Sun.

The Sun's progress across the heavens by day was one of the earliest methods used for marking time. The Romans divided the daylight into eight portions called hours. In winter the hours were short and in summer they were long. A simple way of reading time from the Sun is to use a sundial. A shadow-stick, called a gnomon, marks out the hours by the position where its shadow lies. The shadow slowly moves all the time as the Sun follows its course through the sky. Before cheap clocks were invented, most people had to rely on the Sun for keeping time.

Today it is much easier to use a watch or clock, of course. Our day is divided into 24 equal hours. Accurate clocks in special laboratories give an exact reading of the time, precise to millionths of a second. But who keeps a check on the master clocks? Astronomers do this by observing the positions of the Sun and stars. If all the world's clocks stopped at once, astronomers could reset them to the correct time.

If everyone took the time directly from the Sun, a serious problem would arise. Sun time readings differ for places with different longitudes. For each 15° of longitude, the Sun time changes by a whole hour. All the people living in one country need to agree on just one time for the whole country if there is not to be a great deal of con-

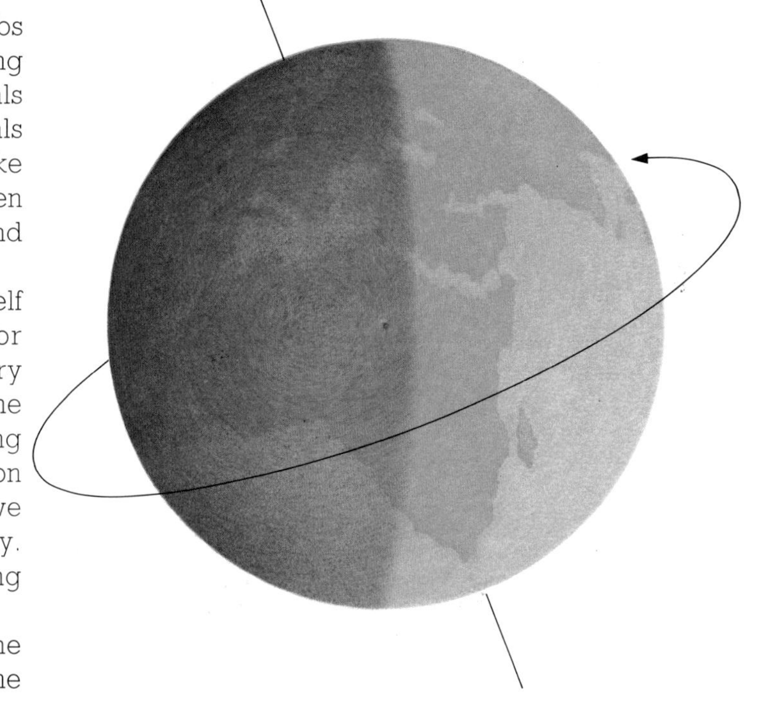

*One half of the Earth is in sunlight while the other half is in darkness. As the Earth spins round, each place experiences day and night.*

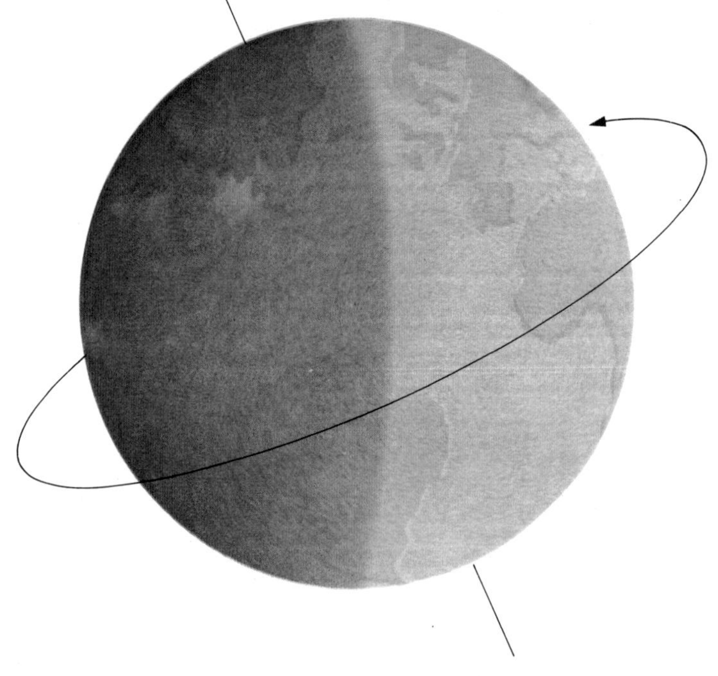

fusion! The time taken directly from the Sun is called local time, whereas the time agreed for a whole country or area is termed civil time.

Civil time is usually chosen to be close to the local time for most of the country. Large countries, such as the United States, need several time zones. The entire surface of the Earth is divided into time zones. Each zone is roughly 15° of longitude, but the boundaries are not straight. Instead, they follow the borders between countries or states. When you travel across a country or state border, you may have to reset the time on your watch. Supersonic airliners can travel faster than the speed of the Earth's rotation. This gives travellers going west the strange sensation of arriving at their destination "before" they started their journey!

Sometimes the governments of countries decide to change the civil time by one or even two hours. When this happens, civil time is no longer close to local time. The time change is done to make better use of daylight so that artificial lights are needed less. Everyone is persuaded to get up an hour earlier because they are made to alter their clocks!

*With a simple sundial you can read the local Sun-time correct to about 15 minutes. This fine dial is in the Great Court of Trinity College, Cambridge.*

*This map shows all the time zones throughout the world. Zones differing from an adjacent zone by less than one hour are indicated in yellow.*

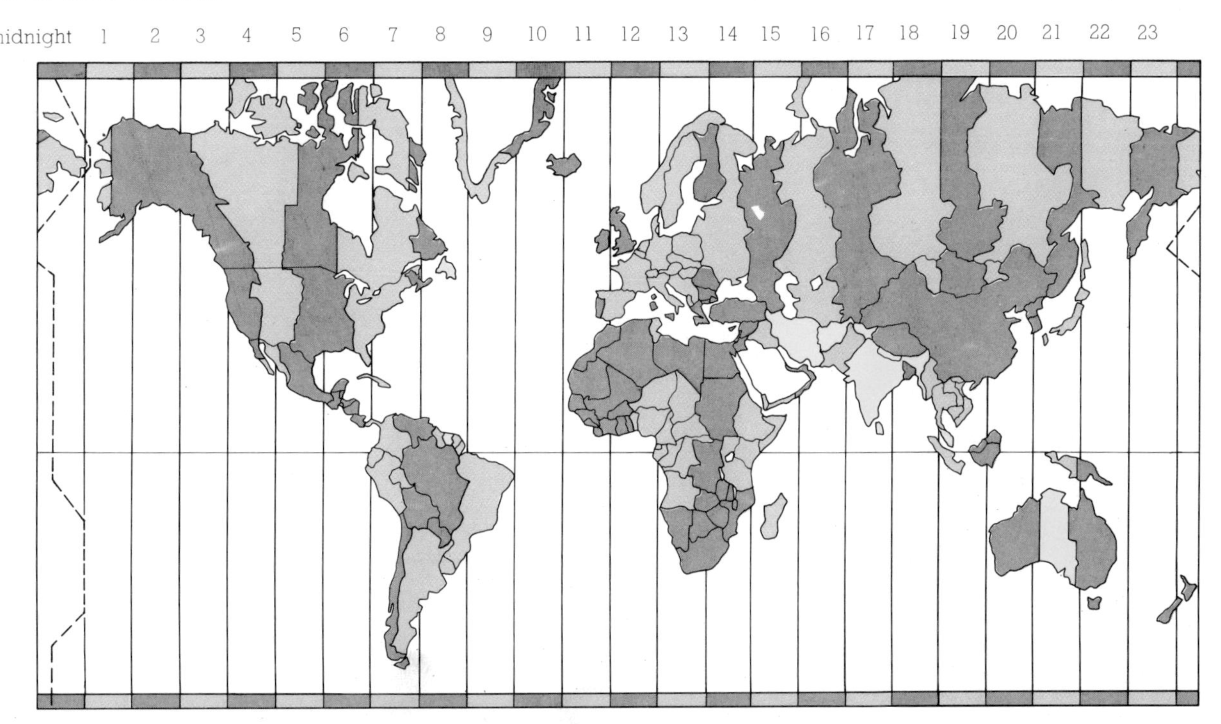

# The Nightly Parade of the

As the Sun sets on a clear day, pin-points of starlight gradually become visible in the darkening sky. The day-time sky has stars in it too, but it is impossible to see them because of the Sun's great brilliance. As the Earth's rotation carries us into darkness, the nightly parade of stars begins. Just as the Sun rises in the east and sets in the west, each star seems to rise and set as the Earth spins round. While some stars sink out of view below the western horizon, new ones are rising in the east. By dawn, the starry sky has changed considerably from its appearance at dusk.

For people living north of the equator, one star hardly moves in the sky: the North Star, or Polaris. This star has, by chance, a rather special position in the sky, for it is almost directly over the Earth's north pole. Imagine the Earth's rotation axis pointing out into space towards the north pole of the sky. The star Polaris is very close to the north pole of the sky. There is also a south pole of the sky in the exact opposite direction in space. Only people who live south of the equator can see it. No bright star lies close to the south pole of the sky.

Some stars close to the pole never rise and set but are always above the horizon. These are called circumpolar stars. As the Earth turns on its axis, the stars trace circles through the sky, around the poles. An interesting way to show the paths the stars sweep out during the night is to take a time exposure photograph. A camera is set pointing at the sky with the shutter open. The photograph shows the circular paths of the stars.

The number of circumpolar stars visible from any particular place depends on its latitude on the Earth. Since the North Star is almost over the Earth's north pole, an observer there sees it right overhead. All the other stars seem to move in circles around the North Star, so none of them ever rises or sets! To an observer on the equator, the North Star is just on his northern horizon. For him no stars are circumpolar. All stars rise and set. If our observer travels north, the North Star will look higher and higher in the sky the further he goes.

Just as the Sun's position can be used to tell the time by day, the stars make a night-time clock. The famous group of stars called the Plough or the Big Dipper is circumpolar

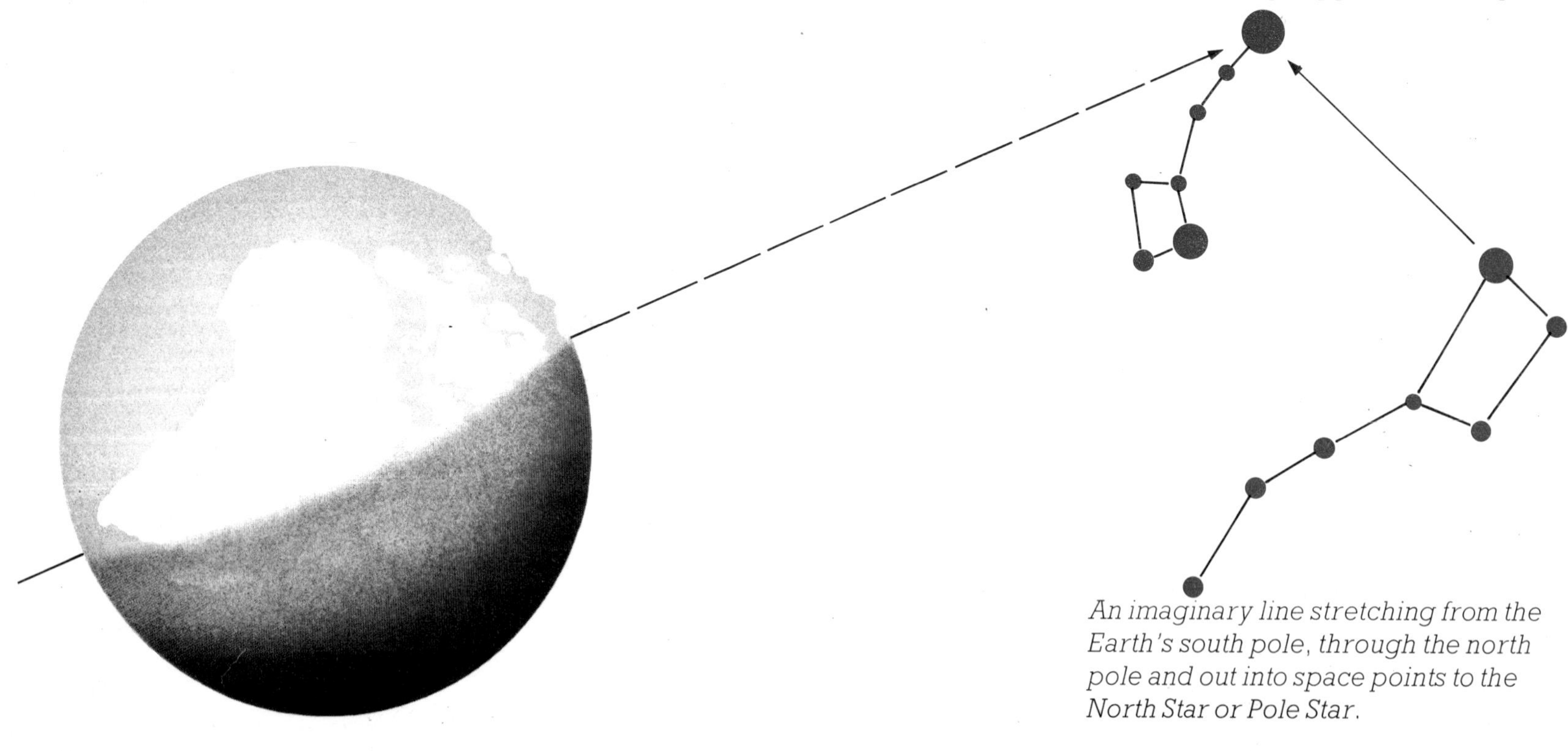

*An imaginary line stretching from the Earth's south pole, through the north pole and out into space points to the North Star or Pole Star.*

# Stars

for people in Europe and the United States. Two of its stars form a line that points to the North Star. This line of stars is like the hour hand of a gigantic 24-hour clock in the sky. During the course of a day, the Big Dipper makes a whole circle around the north pole of the sky.

*A time exposure photograph shows the circles swept out on the sky as Earth spins beneath the stars. This is the southern sky seen from Australia.*

*The North Star does not seem to move during the night. However, its height in the sky depends on latitude. Within the Arctic Circle, the North Star is nearly overhead. Near the equator it is close to the horizon.*

# The Earth's Yearly Journey

The planet Earth is a member of the solar system. It is just one in a family of nine major planets circling the Sun. Each year the Earth travels one complete circuit of its orbit around the Sun. The Earth and everything on it is speeding along at 100,000 kilometers per hour on the round trip. We do not feel any sensation of the high-speed motion because we keep moving smoothly without any jerks.

What keeps the Earth on its path round the Sun? Although the Sun is 150,000,000 kilometers away from the Earth, there is a strong gravitational pull between the two bodies. The force of gravity attracts the Earth towards the Sun. However, because the Earth is moving, it does not fall in towards the Sun. Instead it circles round the Sun. The Earth's orbit is very nearly a circle, but not quite. It is slightly elongated and the Sun is nearer to one side of Earth's orbit. The Earth is 5,000,000 kilometers nearer to the Sun in January, when we get closest, than it is in July, at the farthest point from the Sun. The difference in distance causes only a very slight change in the warmth of the Sun's rays and is much smaller than the familiar seasonal changes.

How can we tell when a year has gone by? The most important changes during the year for the plants and crea-

*The Earth is held in its path round the Sun by the force of gravity. The spin axis is tilted at about 67° to the orbit. This tilt causes the seasonal changes.*

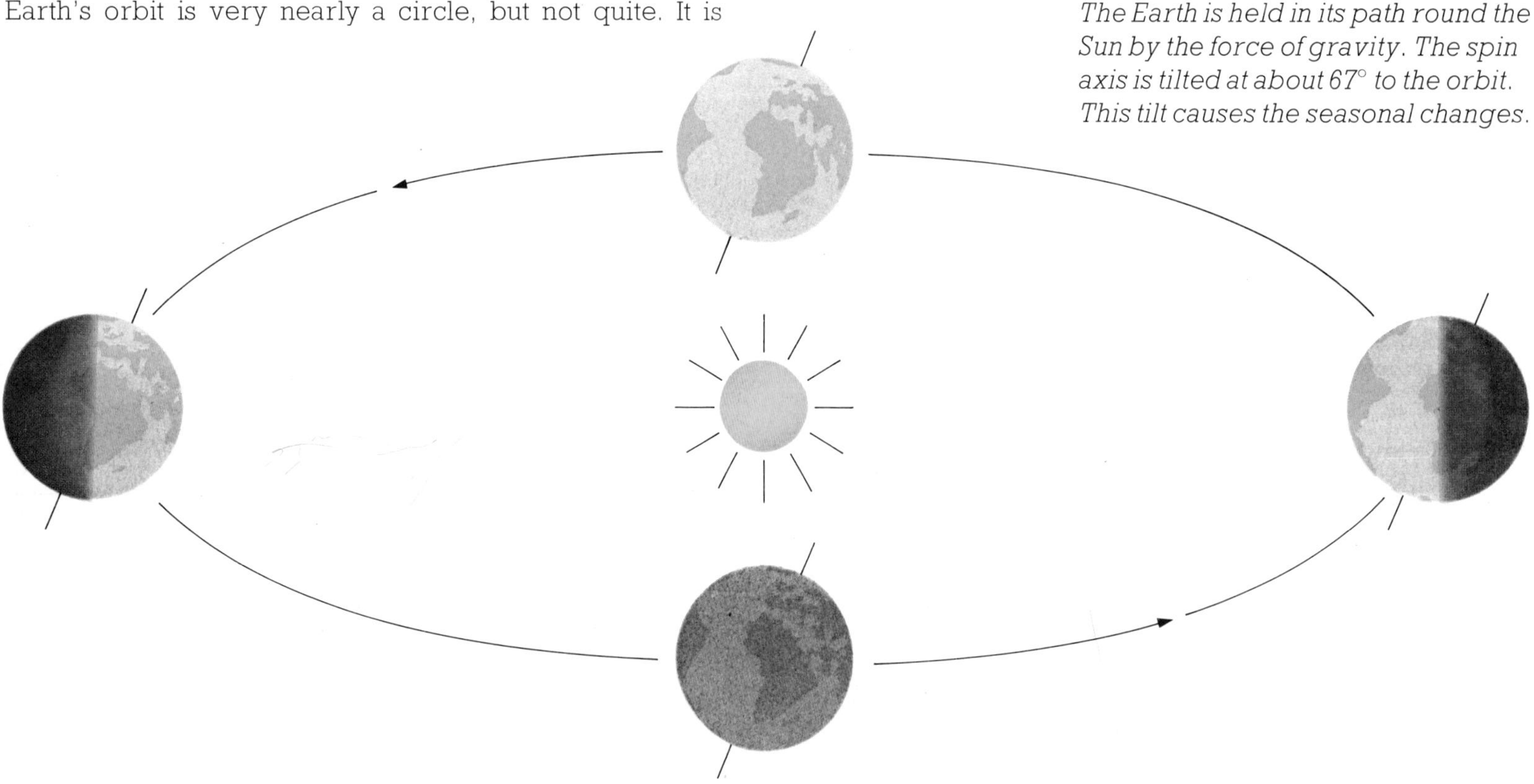

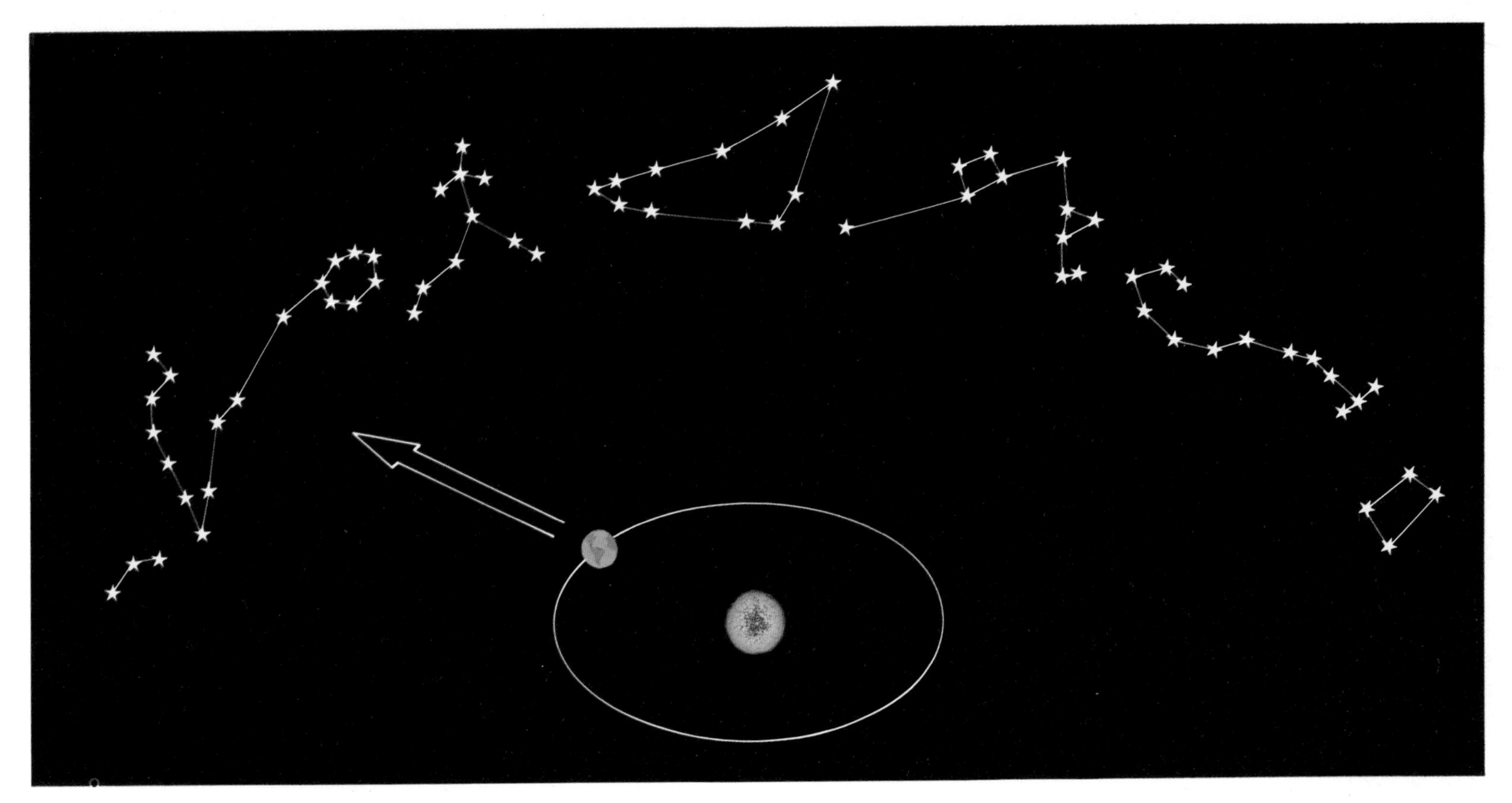

*The Earth takes a year to complete one circuit of its orbit round the Sun. At different times of the year the night-time side of Earth looks out at the stars in different directions in space. From left to right, the constellations are Aries, Pisces, Aquarius, Capricornus, Sagittarius, Scorpius, Libra.*

tures living on the Earth are the changes of weather that come with the seasons. People, particularly, need to keep track of the seasons so that crops can be planted at the right times. Plants and animals have adapted their behavior to fit in with the seasons. Some animals hibernate during the cold winter. Most plants grow when there is plenty of warm sunlight and remain dormant in winter.

For astronomers, there is another important change during the course of the year. The stars visible at night are different from season to season. Each night, the stars rise above the horizon four minutes earlier than they did the previous night. After an interval of fifteen days, the stars are rising and setting a whole hour earlier. Over the space of six months, the stars' rising times change by 12 hours. Consequently, the night sky of winter looks completely different from the pattern of stars visible in summer.

Here is an easy way to understand how the seasonal changes in the night sky happen. Think of the Earth at a certain place in its orbit, in December, for example. The night-time side of the Earth faces a particular direction in space, so the stars in that part of space are visible at night. After six months have passed, the Earth has travelled to the opposite side of the Sun. The stars that could be seen at night in December are now behind the Sun. They cannot be seen because they are only above the horizon during the day. The night-time side of the Earth now faces the opposite direction in space. The night sky of July displays a different set of stars.

When a whole year has passed, the same familiar stars once again become visible. Although the weather may be quite different from year to year, the changes in the sky are the same. The positions of the stars tell us where the Earth is in its journey and help us to make the calendar.

# The Seasons and the Calendar

As summer changes to autumn and winter, the days get shorter and the nights longer. Each day the Sun climbs a little less high in the sky. With the onset of winter in the northern hemisphere, it rises and sets further south. These changes occur because the Earth's rotation axis is tilted at an angle to its orbit around the Sun. If the Earth's axis were at right angles to its orbit, there would be no seasons. In fact, it is slanted by $23\frac{1}{2}°$ and so we get the seasonal changes.

When summer comes to the northern half of the Earth, the north pole tilts towards the Sun. The Sun gets overhead at more northerly latitudes each day until June 21, midsummer's day. On this day, which is properly called the summer solstice, the Sun just gets overhead at the Tropic of Cancer, latitude $23\frac{1}{2}°$N. The summer solstice is the longest day.

Places inside the Arctic Circle (latitudes $66\frac{1}{2}°$ to 90°) experience some days when the Sun never sets! These places are tilted over to face the Sun even at midnight. This is how countries in the far north became called "the Land of the Midnight Sun". The northernmost city of the United States, Fairbanks in Alaska, lies just south of the Arctic Circle. In the summer it is light enough for outdoor sports at midnight.

In winter, however, the long hours of summer daylight have to be paid for. The Sun just manages to crawl 2° above the horizon for four or five hours. The housebound people look forward to the lengthening days as they go about their business by artificial light.

While the far north endures the long winter nights, the southern hemisphere enjoys its summer. The Sun has moved southward in the sky. At the autumn equinox it was overhead at the equator at midday. The north pole is now tilted away from the Sun, and the south pole towards it. The winter solstice for the north occurs on the same day as the summer solstice for the south. The Sun is overhead at midday at latitude $23\frac{1}{2}°$S, called the Tropic of Capricorn. When the next equinox comes in March, the north experiences spring, while autumn comes to the south, and the whole cycle starts once again.

*At Stonehenge, in southern England, the Sun rises precisely above a marker stone outside the main circle of stones on midsummer day. Four thousand years ago this monument was used to find midsummer and midwinter before the invention of writing and the calendar.*

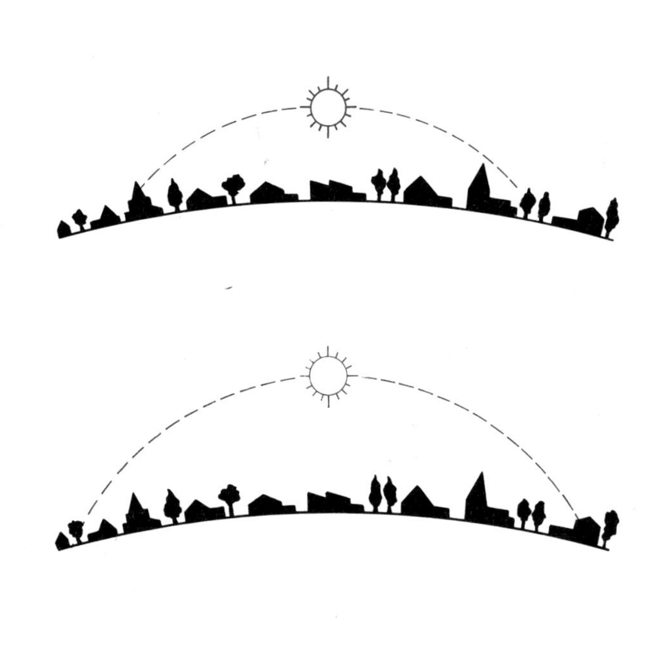

We keep track of the year by using a calendar. We need to have a calendar that everyone uses so that appointments can be made and kept. We also need to keep the months in step with the seasons. In the past great problems arose when the calendar got out of step with the seasons. For example, the best month for sowing seeds gradually changed over thousands of years. By observing the stars, we now know that a year does not contain a whole number of days. During one circuit of the Sun, the Earth spins on its own axis 365 times plus a further quarter of a turn. So, there are very nearly $365\frac{1}{4}$ days in a year. To take account of the quarter of a day, we have a year of 365 days, with an extra day added in leap years. Leap years are the years that can be divided exactly by 4, such as 1980, 1984, etc. However, years ending in 00, like 1900, do not count as leap years unless the year is exactly divisible by 400. We need to take out an extra day per century to remain in step with the seasons.

*The Sun's height in the sky and its rising and setting points on the horizon change with the seasons. The Sun only rises exactly in the east at the equinoxes.*

*The changing aspect of the night sky: Orion is shown at the same time in the evening at intervals of one month.*

# The Constellations and

For as long as people have observed the night sky, they have imagined the shapes of creatures and objects in the patterns of stars. These star patterns are called constellations. Some well-known constellations are easy to pick out. Others are made up of faint stars that are hard to see except in very dark places away from town lights. Some of the constellation names used now are thousands of years old, while others are comparatively modern. The southern half of the sky has mostly modern constellation names because civilisation did not reach the southern hemisphere until explorers from Europe travelled there.

The older constellations feature some of the characters from Greek mythology. They include the hunter Orion with his dog Sirius, Perseus and the princess Andromeda, whom he rescued from a monster. Andromeda's mother, Cassiopeia, and her father, Cepheus, are there in the sky too. There are many stories to be told in the stars.

For astronomers today the word "constellation" has a special meaning. The sky is divided up into 88 areas, in much the same way that a country is divided up into states or counties. Each of these areas is called by the popular constellation name for the bright stars within it. All the

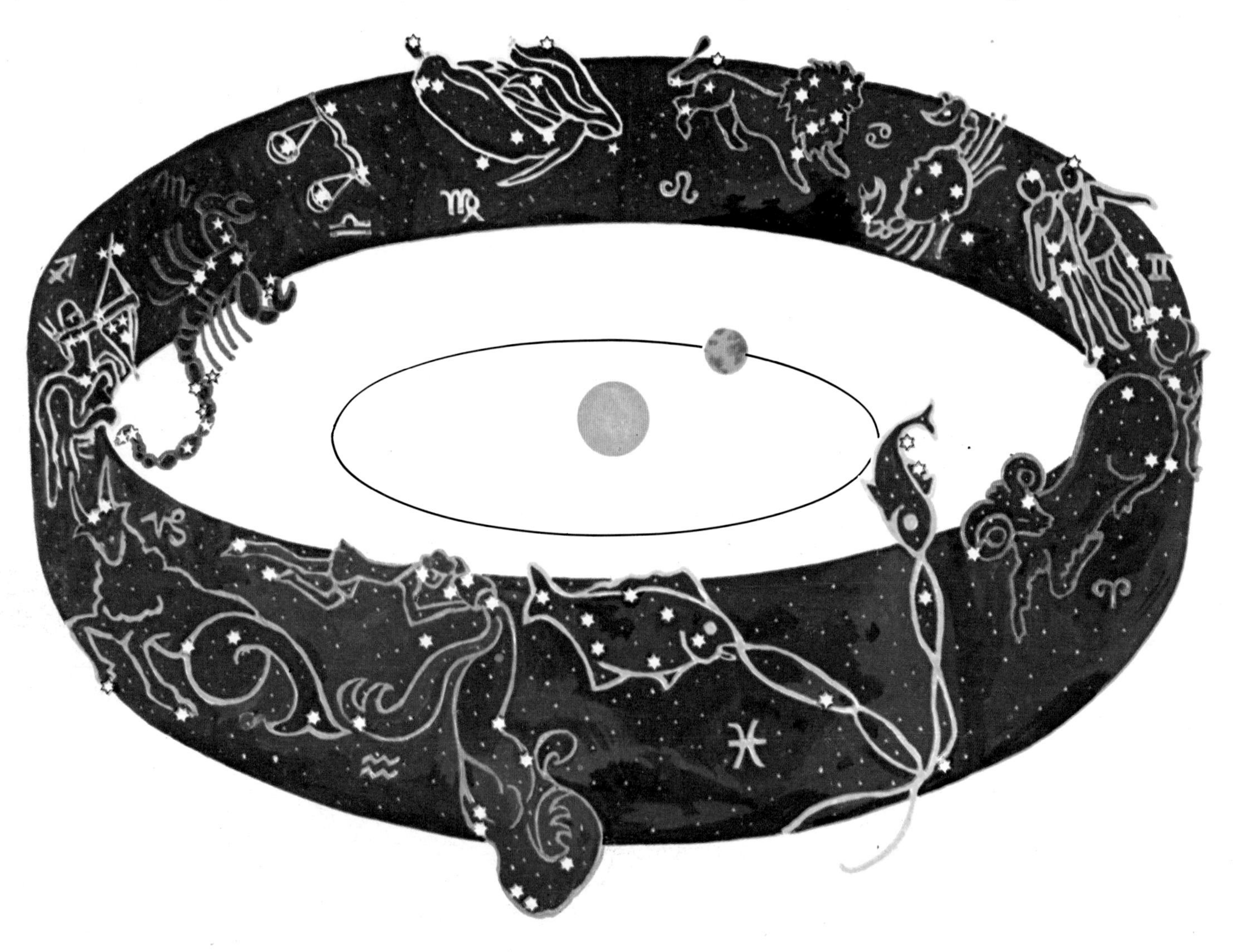

# the Zodiac

stars, even the faint ones, inside the boundaries of a constellation belong to it. The official names for the constellations are in Latin and these names are understood by astronomers of all nationalities.

Twelve of the old constellations are well known to most people because they belong to the zodiac. The zodiac is a band of constellations circling our entire sky. This band is important because the paths of the Sun, the Moon and the planets are always within it. If you picture the Earth in its orbit around the Sun, the stars of the zodiac constellations seem to encircle it. The paths of all the planets around the Sun, and the orbit of the Moon around the Earth, all lie very nearly in the same plane. The shape of the solar system is a large, thin, circular disk. As a consequence, most members of the solar system are always found in our skies against the background of the zodiac constellations.

Astrologers believe that the exact positions of the Sun, Moon and planets against the background of the zodiac have an important bearing on events on Earth. These positions at the moment of a person's birth form his or her horoscope. Few astronomers believe that there is any truth in astrology. However, some scientists think that the time of year when a person is born may affect his character and general behavior.

Your zodiac sign is the constellation in which the Sun lay when you were born. Although the stars cannot be seen during the day, we know that the stars are really still there. As the Earth orbits the Sun in the course of a year, the position of the Sun as viewed from the Earth seems to move through the twelve constellations of the zodiac. Astrologers divide the zodiac into twelve equal parts, though the astronomers' constellations are not all the same size. So, from the astrologers' point of view, the Sun spends just one month in each sign.

*The twelve signs of the zodiac are a series of constellations in a band circling the solar system. During the year, the Sun's path through the heavens passes through each constellation of the zodiac in turn.*

| | | |
|---|---|---|
| ♈ | **Aries** | *22 March–20 April* |
| ♉ | **Taurus** | *21 April–21 May* |
| ♊ | **Gemini** | *22 May–22 June* |
| ♋ | **Cancer** | *23 June–23 July* |
| ♌ | **Leo** | *24 July–23 August* |
| ♍ | **Virgo** | *24 August–23 September* |
| ♎ | **Libra** | *24 September–23 October* |
| ♏ | **Scorpius** | *24 October–22 November* |
| ♐ | **Sagittarius** | *23 November–22 December* |
| ♑ | **Capricornus** | *23 December–19 January* |
| ♒ | **Aquarius** | *20 January–19 February* |
| ♓ | **Pisces** | *20 February–21 March* |

*The signs of the zodiac used in astrology and the symbols used to represent them.*

When we look at the star patterns in the sky, there is no way to tell immediately whether the stars are all at the same distance from us or whether they are scattered through space. We know that the stars are at a great distance compared with the Sun, Moon and planets or the patterns would change as the Earth orbits the Sun. Ancient astronomers thought that the stars were all fixed on a great, distant sphere. Modern astronomers have many special ways of finding the distances to the stars. They have instruments to detect the smallest change in a star's position. They can measure changes in the light from stars. Measurements like these allow them to calculate star distances. The stars are really scattered through space at different distances. The star patterns we see from Earth are a result of where we happen to be in space. If you were able to travel far out into space to a planet orbiting another star, the skies would have quite a different appearance.

# Star Maps for Autumn and Winter

The star maps and charts on these and the next few pages will help you to find and identify some of the constellations. Constellations that contain only fainter stars have been omitted to make the maps easier to follow.

A set of maps is needed because the sky is changing from hour to hour, and through the course of the year. Each map is labelled with the dates and times for its correct use. Each map consists of two semicircles, one labelled north, and one south. Choose the map for the date and time nearest to the time when you wish to observe. If you stand facing due north, the northern map will represent the brightest stars you can see in front of you. If you now turn to face due south, the other semicircle will identify the stars you can see.

The stars directly overhead are hard to identify from these maps, but once you have begun to find your way about the sky from the easier constellations, you will be able to use more detailed maps to identify the fainter ones. It is a good idea to learn the constellations one by one. Each time you go out, look for those you have already learned and try to find a new one. That way you will soon become an expert on finding your way round the sky.

The stars visible depend on the observer's latitude on the Earth. The maps can be used for a range of latitudes, but the horizon will vary. The approximate cut-off by the horizon at different latitudes is shown by the small marks at the sides of the maps. The times given in the captions are all Greenwich Mean Time.

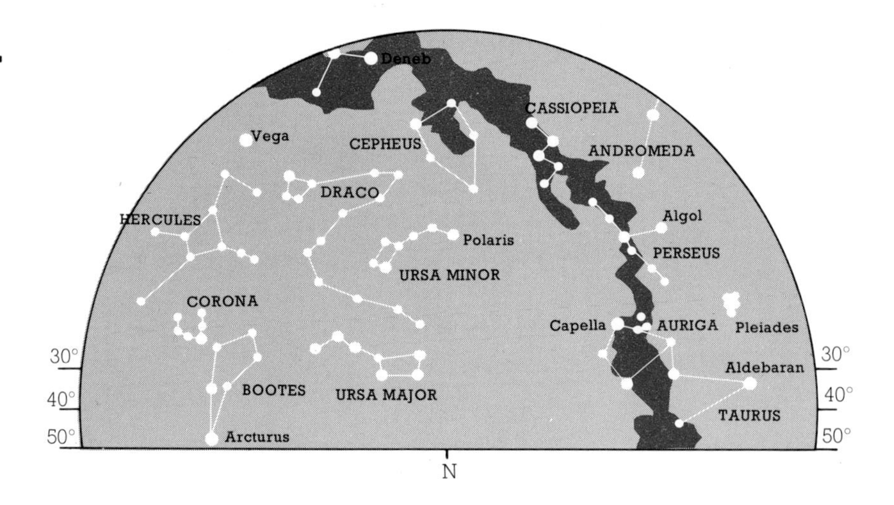


*1 August 1 am*
*1 September 11 pm*
*1 October 9 pm*
*1 November 7 pm*

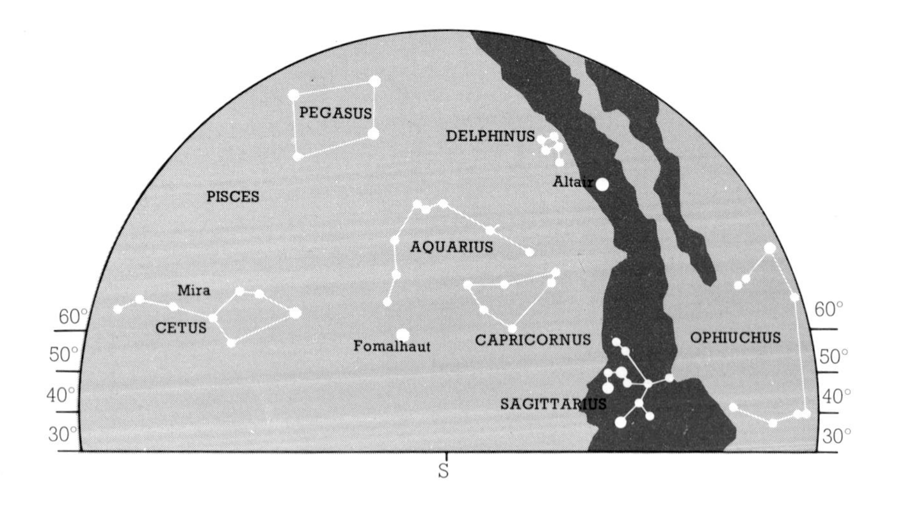


*1 August 1 am*
*1 September 11 pm*
*1 October 9 pm*
*1 November 7 pm*

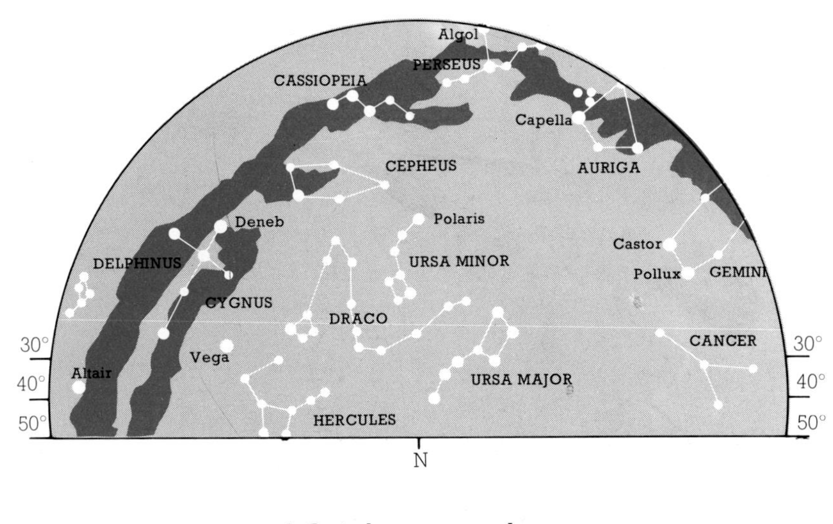

1 October 1 am
1 November 11 pm
1 December 9 pm
1 January 7 pm

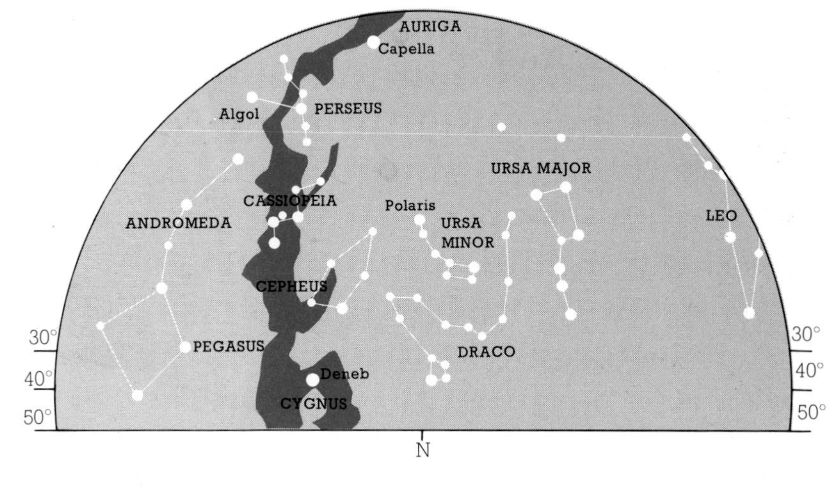


1 December 1 am
1 January 11 pm
1 February 9 pm

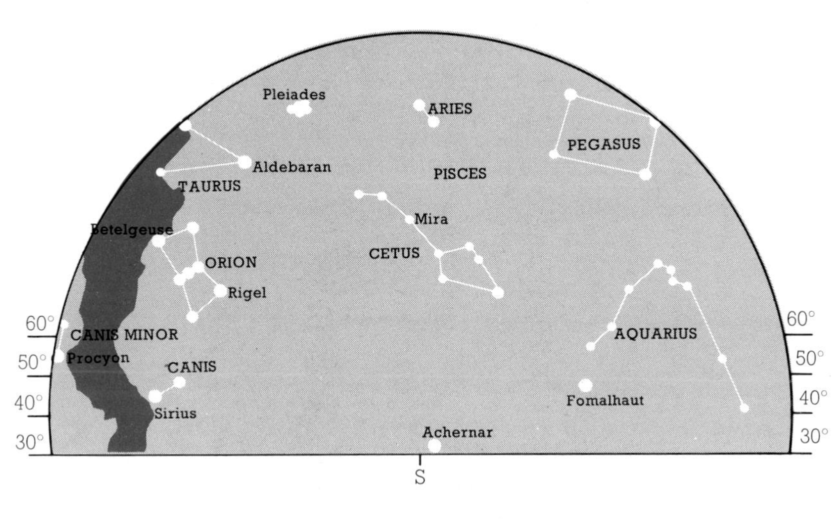


1 October 1 am
1 November 11 pm
1 December 9 pm
1 January 7 pm

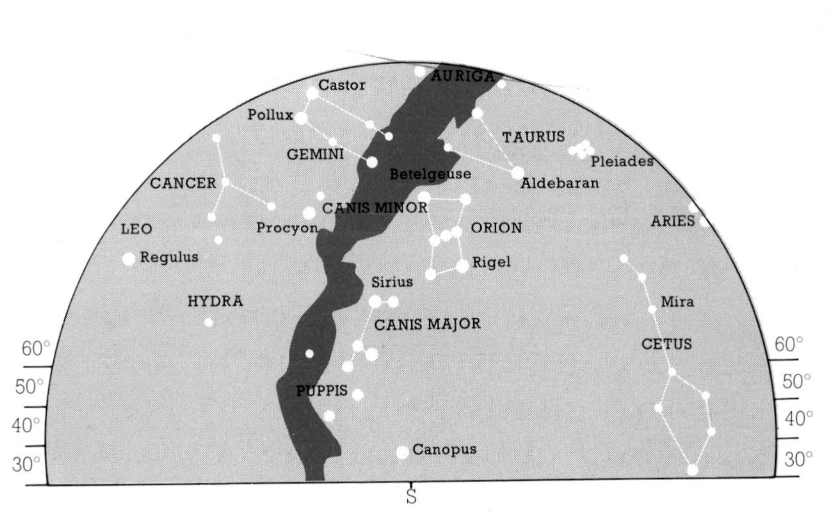


1 December 1 am
1 January 11 pm
1 February 9 pm

# Constellations of

A good starting point for recognizing stars in autumn is the constellation Cassiopeia. The five main stars make the very obvious shape of the letter W, even though none of them reaches the first magnitude of brightness. This starry W is high in the sky during autumn evenings. Observers located at latitudes around 50° to 60° north see it right overhead.

Two stars in the W of Cassiopeia point to another character in mythology, Cassiopeia's husband, Cepheus. And on the other side of Cassiopeia is their daughter, Andromeda. The W is also a signpost to the large constellation of Pegasus, the winged horse. The principal stars in this constellation form a large square known as the Square of Pegasus. When you look for this square, bear in mind that it is large and that none of the stars are first magnitude. On a really dark night see how many stars you can count inside the square. To see a dozen is a good effort, but really keen-sighted people can glimpse more.

Andromeda includes an object of unique interest. The great spiral galaxy M31 is just visible to the eye as a hazy patch of light. Two faint stars in Andromeda lead you to the galaxy M31. This object is a galaxy of stars far beyond the edge of the Milky Way. Its light has taken two million years to reach your eyes. The Andromeda Galaxy is the most distant object visible to unaided human eyes. In a small telescope M31 is a soft glow of light.

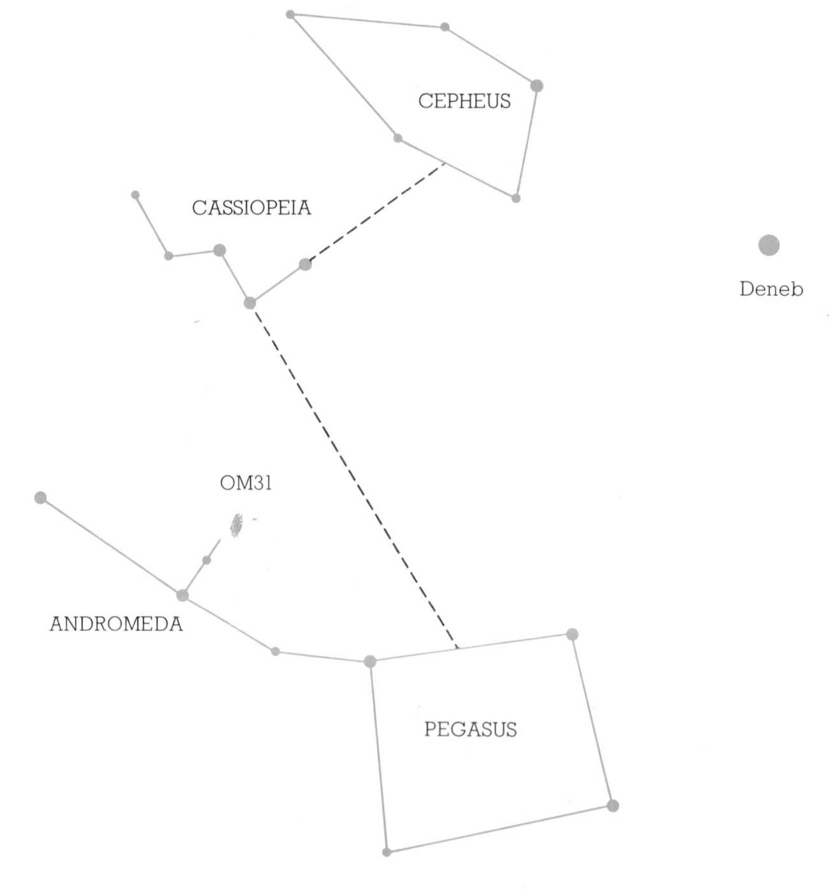

# Autumn and Winter

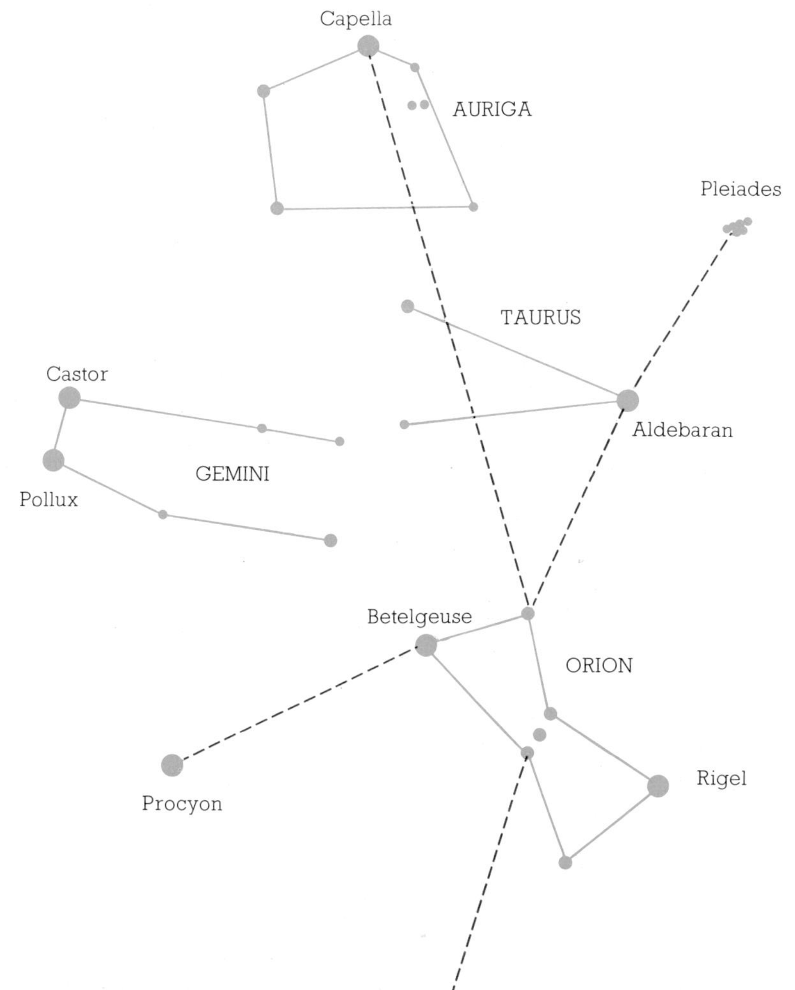


The bright constellations of winter are the best known, as well as being simple to learn. This is a good time to get started on constellation spotting.

One of the most splendid views is that to the south on a January evening. Ahead is Orion, marching across the sky, with three stars forming a neat sword-belt for this hunter from mythology. In Orion seven stars of first and second magnitude make up a memorable pattern. Betelgeuse, the star at upper left, is distinctly red. It contrasts well with Rigel, at lower right, which sparkles blue-white. Orion's belt points to twinkling Sirius in Canis Major, the Big Dog. Sirius is the brightest star visible from northern latitudes. Procyon is another first magnitude star forming a triangle with Betelgeuse and Sirius. Procyon belongs to the Little Dog, Canis Minor.

Gazing above Orion, we find Taurus, the Bull and Gemini, the Twins. Aldebaran in Taurus is red, like Betelgeuse. Near to the Bull's head lies the conspicuous cluster of stars called the Pleiades, or Seven Sisters. Six or seven stars can be seen by the unaided eye, but a small telescope or binoculars will bring dozens more into view.

Higher still in the sky lies Auriga, the Charioteer. This constellation contains the yellowish first magnitude star, Capella.

# Star Maps for Spring and Summer

The stars we can see with the naked eye cover a range of brightness. The brightness of a star as it is seen by eye is called its apparent visual magnitude, or just magnitude. On the magnitude scale, the smallest numbers go to the brightest stars. The most brilliant stars in the night sky are those of zero and first magnitudes. On a dark, moonless night, the faintest stars that can be seen are sixth magnitude. On these maps, three symbols are used to distinguish between the brightest, medium and fainter naked-eye stars.

Fainter stars are only noticeable when the sky is really dark. The atmosphere picks up and scatters any bright light so the sky background is faintly illuminated. In or near a town where there are many street lights, the sky never gets really dark, though it may still be possible to identify some of the brighter constellations. If the Moon is up too, especially if it is nearly full, the sky will be too bright to see more than a handful of stars.

When you have been outside in the dark for a few minutes, your eyes start to get dark-adapted. The pupil of the eye opens up to let in more light and you will find that you can see fainter stars. Once you have become dark-adapted it is best to avoid going back into a brightly lit place until you have finished observing. For looking at the star maps use a dim red torch. Red light will not affect your dark-adaption very much.

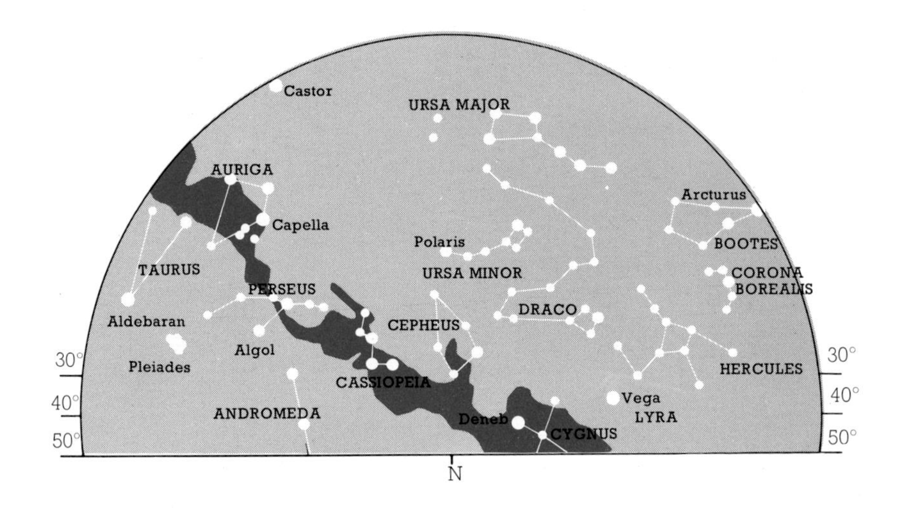


*1 February 1 am*
*1 March 11 pm*
*1 April 9 pm*

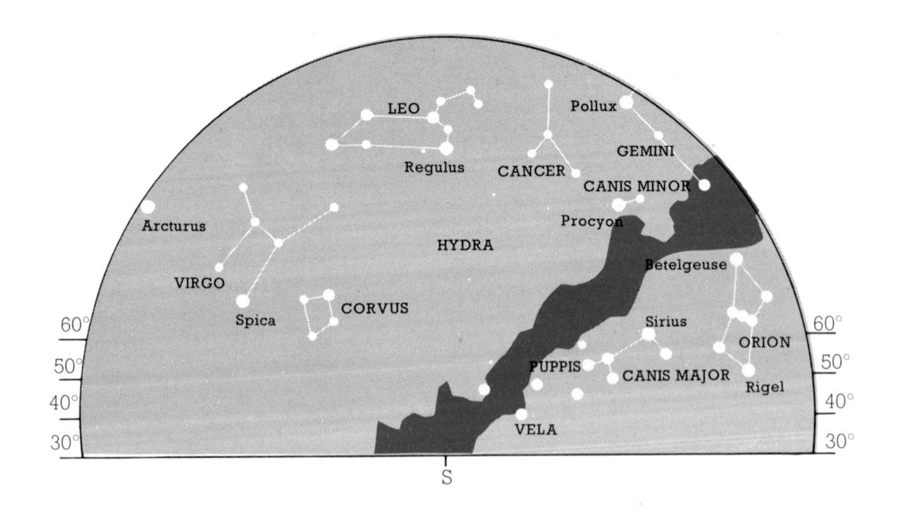


*1 February 1 am*
*1 March 11 pm*
*1 April 9 pm*

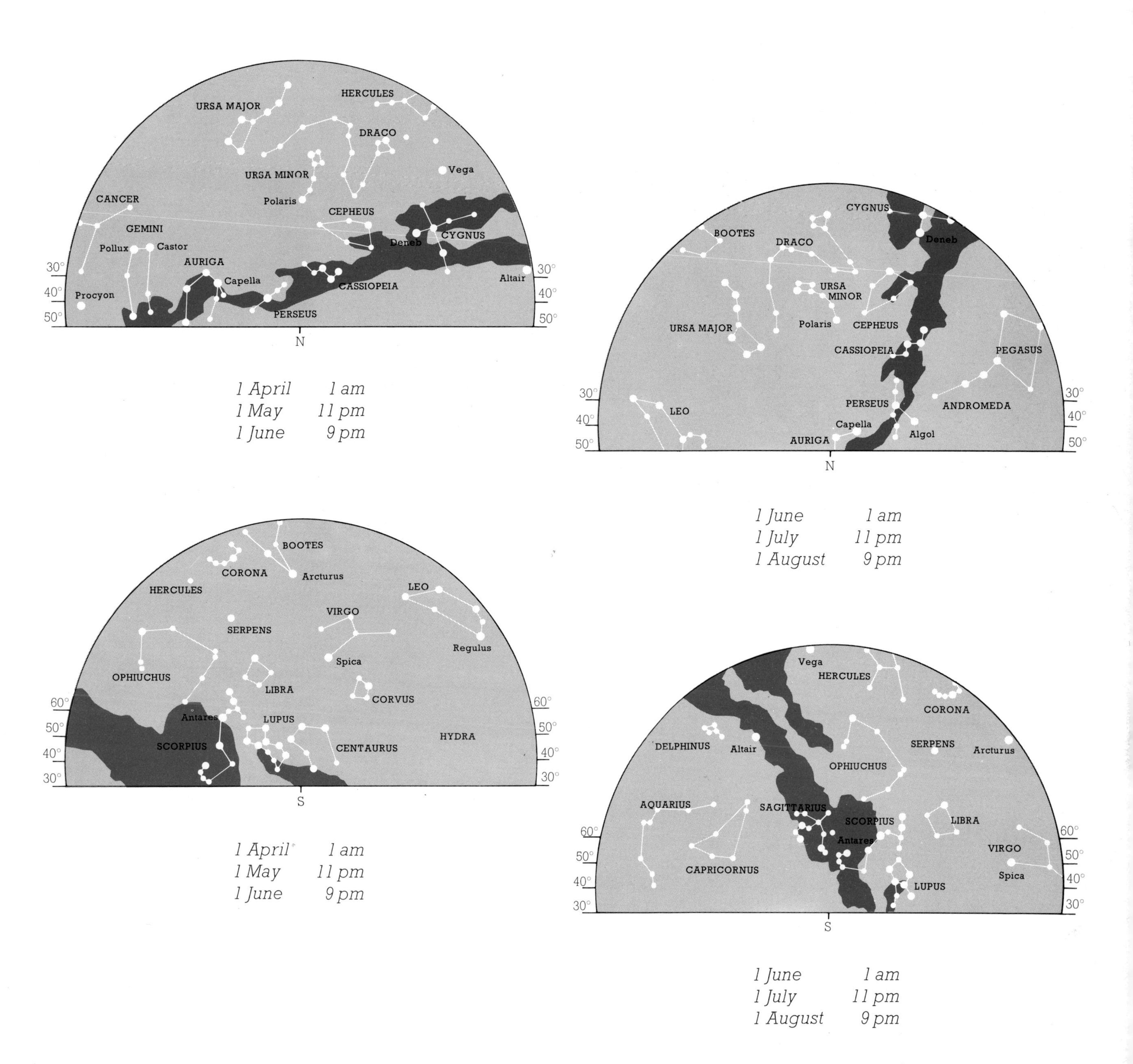


*1 April 1 am*
*1 May 11 pm*
*1 June 9 pm*

*1 June 1 am*
*1 July 11 pm*
*1 August 9 pm*

*1 April 1 am*
*1 May 11 pm*
*1 June 9 pm*

*1 June 1 am*
*1 July 11 pm*
*1 August 9 pm*

# Constellations of Spring

On April evenings the constellation Ursa Major reaches its highest point in the sky. Observers between latitudes 50° and 60° north see it directly overhead. A section of Ursa Major, the Plough or Big Dipper, can be used to find three of the first magnitude stars in spring skies in the following manner.

Following round the curve of the Bear's tail (handle of the Dipper) leads on to Arcturus. Arcturus is one of the brightest stars in the northern part of the sky and it looks reddish. It belongs to the constellations of Bootes, the Herdsman. The other stars in this group are all considerably fainter than Arcturus. Next to Bootes is the small neat constellation of Corona, the Crown. Corona is composed of a semicircle of moderately faint stars. Its distinctive shape stands out well on a dark night.

Continuing to sweep in a smooth arc from the Bear, through Arcturus, leads you on to Spica. This white star, the brightest in the constellation Virgo, is also first magnitude.

One of the most conspicuous constellations of spring is Leo, the Lion. It can be seen by facing south at this time of year. A line through the bowl of the Dipper (through the ploughshare) pointing away from the North Star, leads to Leo. The lion's head makes a shape that is similar to a back-to-front question mark (⸮) or sickle, Leo includes the first magnitude star Regulus.

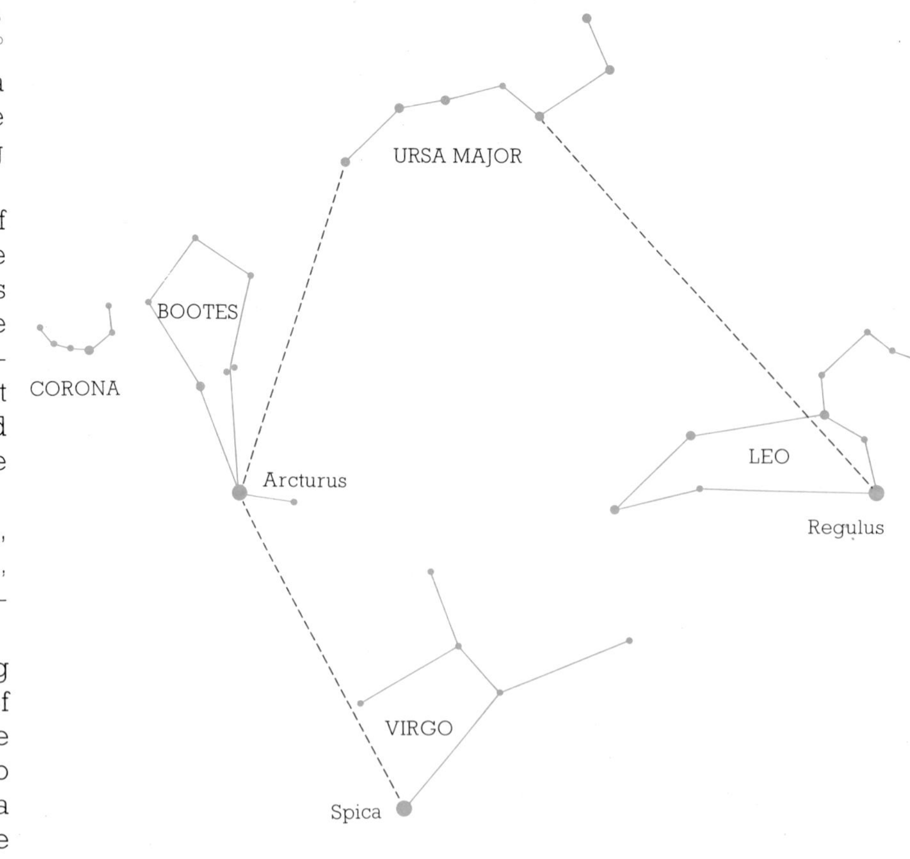

# and Summer

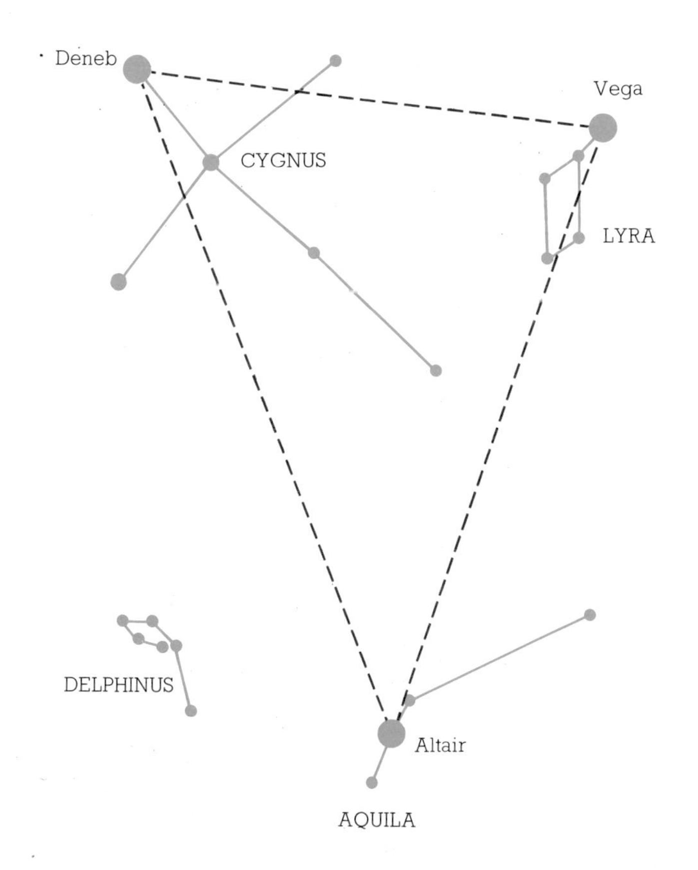


In the skies of the summer evenings three first magnitude stars stand out: Deneb in Cygnus, Vega in Lyra, and Altair in Aquila. For an observer who faces south, these three map out the prominent Summer Triangle.

Deneb is the brightest star in the constellation of Cygnus, the Swan. This group of stars is sometimes called the Northern Cross. It lies right within the Milky Way. Binoculars or a small telescope can be swept slowly around this part of the sky. Rich star fields, many of them thousands of light years away will come into view. They make up the soft glow of light from the Milky Way itself. Close to Cygnus there are dark patches where dust clouds in deep space cut out the faint background of light from distant stars.

Vega is a member of the small constellation of Lyra, the Lyre, and Altair belongs to Aquila, the Eagle. The other stars in these two constellations are much fainter than Vega and Altair. The attractive grouping of stars making up Delphinus, the Dolphin, lies close to Aquila. Although the member stars of Delphinus are all quite faint they form a compact group that is easy to find and to remember.

Fewer bright constellations are to be seen in summer than in winter. In fact, some parts of the summer sky are surprisingly empty to the naked eye. For this reason, and also because the skies darken later in the evening in the summer, it's easier to learn about the sky in winter.

# The Sun's Family

The first astronomers, long ago, noticed that there are five special "stars" that gradually move through the constellations. They became known as the "wanderers" or planets. Planets shine with a steady light, but real stars often twinkle.

Planets are not like stars at all. Our Sun is a typical star. It radiates heat and light of its own, but the planets shine only by the light they reflect from the Sun. Most stars are much larger than planets. Our Sun is a thousand times more massive than the giant planet, Jupiter. The twinkling stars are other suns, much further away from us than any planet.

All the planets visible in the night sky are members of the Sun's family, or solar system. The five planets that can be seen without a telescope are Mercury, Venus, Mars, Jupiter and Saturn. Mercury is closest to the Sun. It is not easy to pick out because it is never far from the Sun in the sky. Venus is also closer to the Sun than Earth. This brilliant planet is seen at its best at dawn or dusk and so it is often called the "morning star" or "evening star". Mars is "the Red Planet", so named because of its color. Jupiter and Saturn, both giant planets, can often be seen shining with steady yellow light. Mars, Jupiter and Saturn lie further from the Sun than the Earth.

After the invention of the telescope, astronomers found three, more distant planets. Uranus was discovered in 1781, Neptune in 1846 and Pluto in 1930. All nine planets travel in orbits around the Sun. They all journey in the same direction. The planets nearest the Sun take the least time. Mercury, nearest to the Sun, makes a circuit in only 88 days. Earth takes a year, and Jupiter 12 years.

Johannes Kepler studied the motion of the planets. In

1609 he discovered that the orbits of the planets are slightly stretched circles, called ellipses. An ellipse has two focal points. For each planetary orbit the Sun is at one of the focuses. This means that the distances of the planets from the Sun change by small amounts as they travel in their orbits.

Kepler found out how the planets move, but it was Isaac Newton who realized that the force of gravity holds the planets in their orbits. The Earth's gravity makes objects that are dropped fall to the ground. If the Sun's gravity did not constantly keep tugging at the planets, they would fly off into the depths of space.

The inner planets have all been visited by spacecraft. In the 1980s the Voyager spacecraft, launched by the USA, will reach the outer giants of the solar system. The two Voyagers will photograph the planets and their many moons in great detail.

The Sun's family has other members apart from planets. Swarming between Mars and Jupiter there are thousands of asteroids or minor planets. Comets with their streaming tails, visit us from the farthest parts of the solar system. Dust is scattered in the space between the planets, as well as stones or meteorites. These space rocks burn up if they crash through Earth's atmosphere, creating a meteor trail or "shooting star".

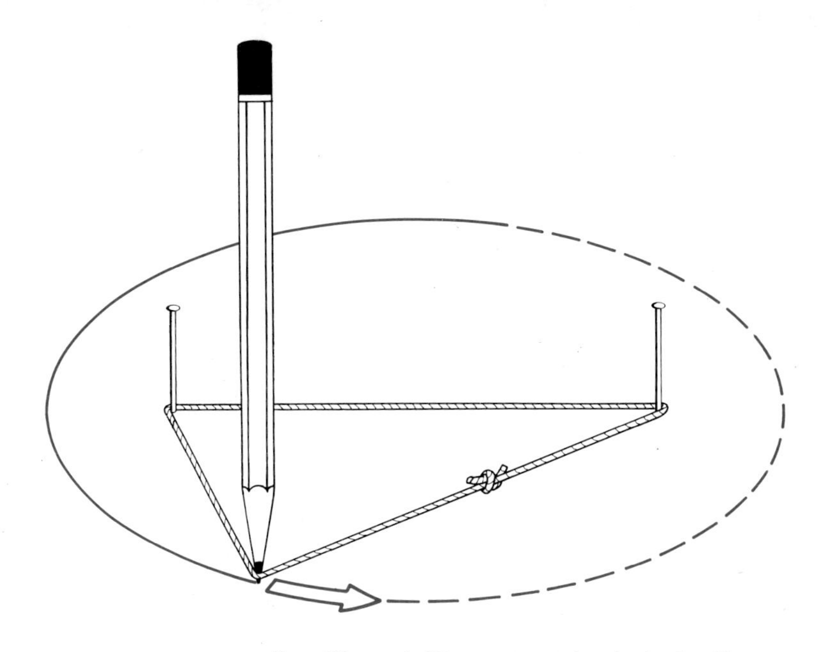

*An ellipse is like a stretched circle. To sketch an ellipse, tie a length of string loosely to two pins fixed on a board. Stretch the string tightly with a pencil and move it around the pins. The pencil traces an ellipse. The two pins mark the focus points of this ellipse.*

Many of the planets have moons orbiting them, rather like miniature solar systems. Jupiter has at least thirteen moons, four of which can be seen in a small telescope. Gravity holds the moons in their orbits around their planets, just as it keeps the whole of the Sun's family together.

*The orbits of the major planets are shown drawn to scale. Between the orbits of Mars and Jupiter lies the broad band of asteroids. The planets are drawn in their correct relative sizes, although they are greatly enlarged compared to the orbits. Notice how small the inner planets and their orbits are compared with the outer giants. All orbits in the planetary system, except for those of comets, are close to one plane.*

| **The Planets** | **Distance from the Sun** | | **Time to orbit the Sun in years** | **Mass compared to the Earth** | **Radius compared to the Earth** |
|---|---|---|---|---|---|
| | Compared to Earth | In millions of kilometers | | | |
| **Mercury** | 0.39 | 58 | 0.24 | 0.06 | 0.38 |
| **Venus** | 0.72 | 108 | 0.62 | 0.82 | 0.95 |
| **Earth** | 1.00 | 150 | 1.00 | 1.00 | 1.00 |
| **Mars** | 1.52 | 228 | 1.88 | 0.11 | 0.53 |
| **Jupiter** | 5.20 | 778 | 11.86 | 318 | 11 |
| **Saturn** | 9.54 | 1430 | 29.46 | 95 | 9 |
| **Uranus** | 19.18 | 2870 | 84 | 15 | 4 |
| **Neptune** | 30.06 | 4500 | 165 | 17 | 4 |
| **Pluto** | 39.44 | 5900 | 248 | 0.1? | 0.5? |

# Eclipse of the Sun

A total eclipse of the Sun must be one of the most eerily beautiful sights in nature. Only a remarkable coincidence makes it possible for us to witness this spectacle. The Sun is a great, luminous ball, 109 times the diameter of the Earth and at a distance of 150 million kilometers. The Moon is only one quarter the size of the Earth, but it is 400 times nearer than the Sun. Of course, things look much smaller when they are at a great distance than when they are close by. The difference between the Sun's and the Moon's distances compensates for their difference in size. As a result, the Sun and the Moon look very nearly the same size in our skies.

It is an amazing coincidence that planet Earth has a satellite which coincides in apparent size to the Sun. Although the Sun is 390 times further from us than the Moon, it is also 390 times larger. The greater diameter is exactly cancelled by our viewing the Sun at a greater distance. Without this beautiful phenomenon we should know very much less about the outer atmosphere of the Sun. No other planet in our solar system can see the eclipses as well as we can.

The Moon takes roughly a month to circuit the Earth in its orbit. Two or three times a year, on average, the Moon's

*During an eclipse the faint outer layers of the Sun, or corona, comes into view.*

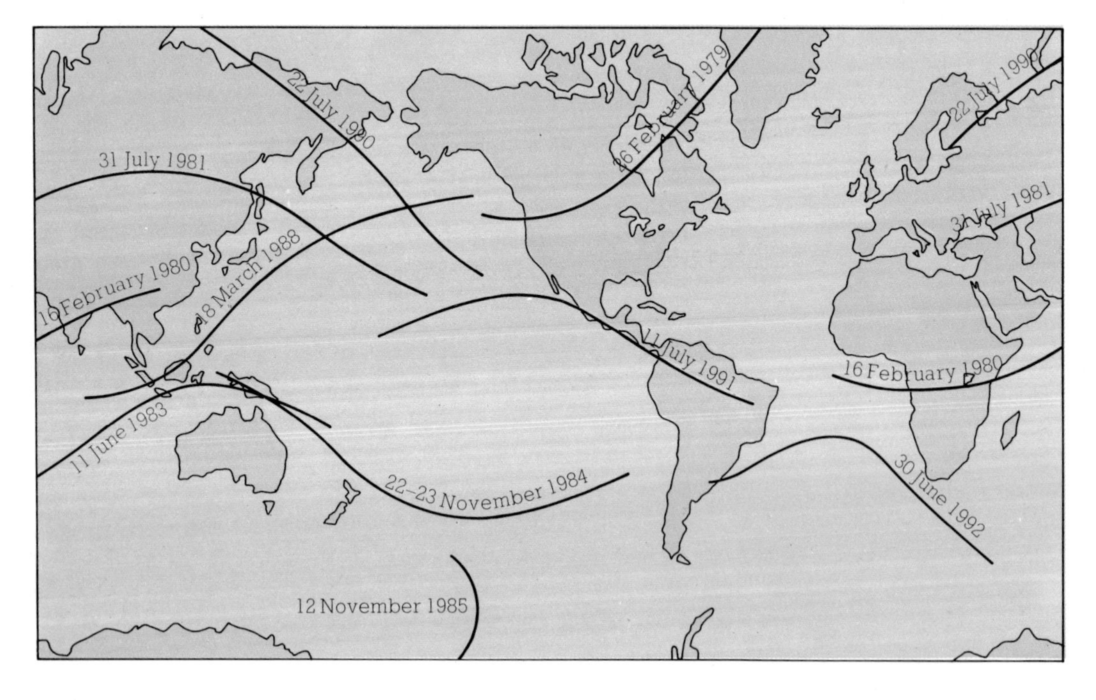



*This map shows the tracks of forthcoming solar eclipses.*

*A series of exposures of the camera over a period of one hour show the progress of an annular eclipse in Costa Rica in 1974. When the Sun is closer than usual and the Moon further away than usual, their sizes do not match. The Moon fails to cut out the Sun completely, as in this case.*

path takes it directly between the Earth and the Sun. At such times an eclipse of the Sun takes place. The dark disk of the Moon blots out all or part of the Sun for a short time. The Moon's shadow measures only a few kilometres across on the Earth's surface, and as it speeds along its orbit the shadow sweeps out a long curved path on the Earth. Any particular eclipse of the Sun will only be seen by people who are somewhere along this long and narrow path. At any one place the totality lasts only two to four minutes. An hour or so before totality is due, the Moon starts to cover part of the Sun. At this stage the eclipse is only partial. In a wide area on each side of the eclipse track only a partial eclipse can be seen.

As the total eclipse becomes imminent, the sky darkens and the stars appear. When the Sun's disk is completely blotted out, a shining white halo shimmers around the black Moon. This is the Sun's corona, a crown of thin hot gas streaming away from the Sun. Close to the black disk of the Moon is a thin ring of reddish gas that is called the solar chromosphere. Sometimes there are prominences visible, as tongues of gas leap up from the Sun.

The edge of the Moon is not perfectly smooth because there are mountains and valleys running along it. Just before or just after totality the Sun may shine through these valleys. This gives the impression of a string of pearls or a brilliantly flashing diamond ring.

Long ago, people feared eclipses. They did not understand what was happening and thought that the Sun might vanish for ever. Today eclipses are closely observed for a different reason. Total eclipses give astronomers rare opportunities to study the fainter parts of the Sun's corona and to look at the chromosphere layer. Long before an eclipse is due, expeditions are carefully planned to places that lie on the totality track. Astronomers try to pick a place that will not be cloudy during the brief eclipse! During the few minutes available, many cameras and instruments are used simultaneously for a variety of experiments. Some teams of researchers take their instruments up in aircraft. That way they can be sure of getting above the clouds, and the plane can chase the Moon's shadow, thus increasing by several minutes the time for which the eclipse is visible.

The distances between the Sun and the Earth, and between the Earth and the Moon are not constant. They vary by small amounts. When the Sun is closer than usual and the Moon is more distant than usual, the Moon looks a little smaller than the Sun. If an eclipse then occurs, the Moon does not cover the Sun completely. Instead a bright ring of sunlight circles the black Moon. Such an event is called an annular eclipse; the word annulus means ring. During an annular eclipse, the sky remains bright and the corona cannot be seen. For those two reasons, annular eclipses are of little scientific value.

# Earth and Moon

No small planet has a moon quite like ours. Mercury and Venus have no moons at all and only two very small chunks of rock hurtle round Mars. In the outer solar system, however, large moons orbit Jupiter and Saturn.

Our Earth and its Moon are like a double planet. The pull of gravity holds them close together. The Moon travels in an orbit round the Earth taking about a month to do so. The word "month" comes from "moon". During the course of a month, we see the Moon go through its regular cycle of phases. Just after new Moon, only a thin crescent can be seen close to the setting Sun. In a week the Moon is half lit and after two weeks there is a full Moon. In another two weeks we are back to new Moon again. These changes happen because the Moon only shines by reflected sunlight, and as it travels around the Earth, we see different amounts of the Moon's sunlit half. At new Moon, the Moon lies between the Earth and the Sun, and the side that is shining faces away from the Earth. At full Moon, the Moon is on the side of the Earth furthest from the Sun and all of the illuminated side is visible.

Two or three times a year, when the Moon goes behind the Earth, it moves into the Earth's shadow. When this happens, the Sun's light cannot reach the Moon because the Earth is in the way, and we say that the Moon is eclipsed. During such an eclipse, the Moon looks dimly red because

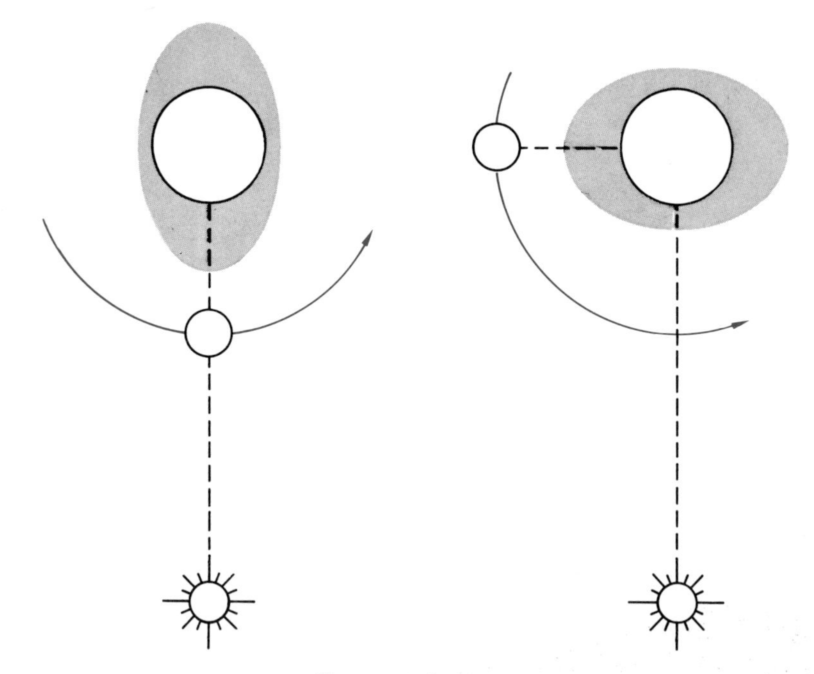

*The gravitational pull of the Moon, and the smaller pull of the Sun combine to cause the ocean tides.*

*The crescent Moon waxes to fullness and then wanes during its monthly cycle.*

*Only the side of the Moon facing the Sun is bright. As the Moon travels round the Earth, different amounts of the sunlit side can be seen.*

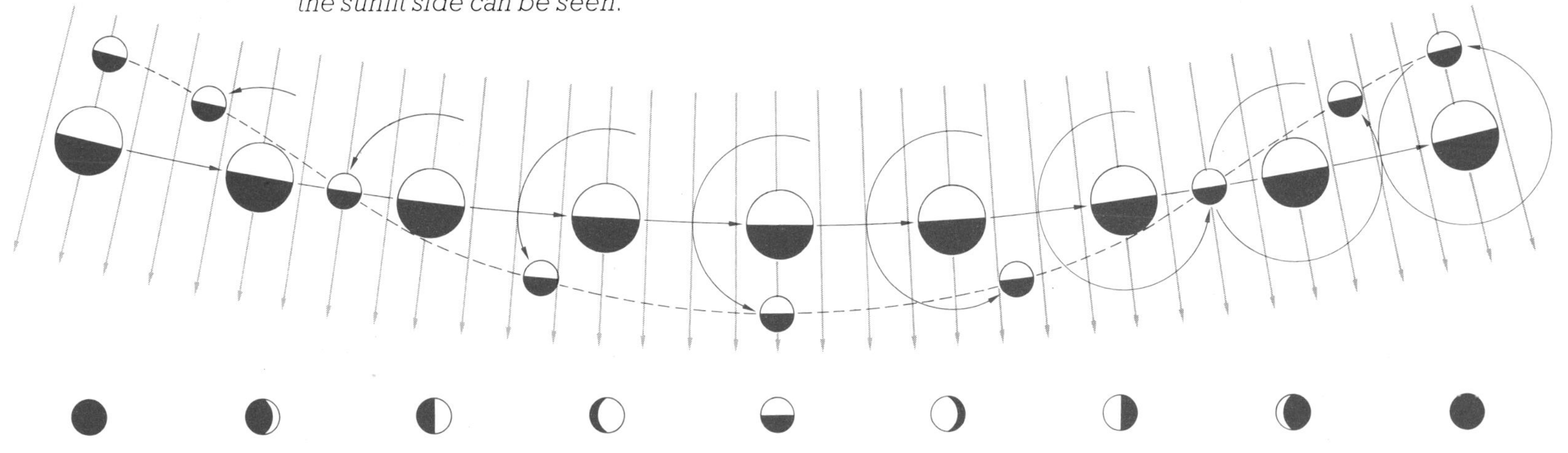

the Earth's atmosphere scatters some sunlight towards the Moon. Eclipses do not take place every month because the Moon's orbit is tilted at an angle to the Earth's path round the Sun. Usually the Moon clears the Earth's shadow by passing above or below it.

You may have noticed that the markings on the Moon never change. The Moon always keeps the same face towards the Earth. Until the first spacecraft were sent to travel round the back of the Moon, nobody knew whether it was similar to the half we can see. Photographs show that the other side is much the same, except that there are fewer dark areas.

More than three-quarters of the Earth's surface is covered by the oceans. The Moon's gravitational pull has the important effect of making ocean tides. Unlike solid rock, liquid water can flow easily. The water surrounding the solid Earth is distorted into a squashed ball under the influence of the Moon's attraction. As the Earth spins on its axis, the bulges in the water seem to sweep round the Earth, causing two tides each day. The Sun, too, has an influence on the tides. When the Moon and the Sun are both pulling on the oceans in line, the highest tides are formed. The tides are delayed approximately 52 minutes each day because the Moon rises about 52 minutes later each day.

# The Great Observatories

Many astronomers work in large observatories where telescopes collect light from the planets, stars and distant galaxies. An observatory has special buildings, called domes, to protect the telescopes from wind, rain and snow. The circular dome has an opening through which the telescope views the sky. At a large observatory there are many telescopes, each designed to assist astronomers in a particular way. Giant telescopes with mirrors four meters or more in diameter look at the faintest stars. Smaller Schmidt telescopes take beautiful photographs of the starry skies.

The major observatories are built high in mountain ranges, above most of the clouds. In the mountains they also escape much of the dazzle from street lights. Sometimes astronomers have to travel thousands of kilometers to reach their telescopes.

Mount Palomar Observatory is the famous American observatory in southern California. It is only 160 kilometers from the blaze of lights in Los Angeles. The newer American telescopes are found in Arizona, on Kitt Peak Mountain. In the nearby city of Tucson the street lights are shaded to protect the telescopes.

*Kitt Peak Mountain in Arizona is now the primary observing site in the USA. Here it is lit by a flash of lightning.*

*At Cambridge, England, several small radio dishes are linked up by cables to make an interferometer. This type of telescope can see detail as finely as a single dish five kilometers across.*

*This radio telescope in West Germany has a dish 100 meters across. It is the world's largest, steerable radio telescope.*

Most of the great observatories are in the northern hemisphere, where most astronomers live. However, American and European astronomers want to know more about the stars and galaxies in the southern skies. For this reason, new observatories have been constructed in South America and Australia.

New observatories located in the clear mountain air of the Chilean Andes are searching the skies of the southern hemisphere. For example, several European countries run an international observatory in Chile. Each of these observatories is equipped with the very latest modern telescopes. One important project they are working on is to make high quality photographs of the southern skies.

British and Australian scientists joined forces to found the Anglo-Australian Observatory in New South Wales. Here they have a telescope with a mirror 3.9 meters in diameter, the fourth largest in the world. The dome for the telescope also contains computers, offices for the researchers, a library, kitchens, and darkrooms for developing photographs of the sky. By day, the astronomers sleep in another building close to the huge dome.

Apart from astronomers, a modern observatory requires people to run computers, to build instruments and to repair the telescopes.

Many galaxies send out radio waves. These are detected by large radio telescopes. Fortunately, clouds do not block out radio waves, so a radio astronomy observatory can be built in a cloudy place. In England there are large radio observatories near to Manchester and at Cambridge. Several large observatories are at work in the USA, USSR and Australia. Radio telescopes can operate during the day as well as at night.

X-rays cannot travel very far through our atmosphere. Space scientists have made automatic astronomy observatories. These orbit Earth, high above the atmosphere. Telescopes controlled by radio signals beamed from the ground examine the invisible X-rays from exotic stars. Eventually, astronomers hope to have orbiting observatories containing several types of telescope. Then the weather will never interfere with practical astronomy.

# Telescopes

For thousands of years, watchers of the skies could only study the stars by eye. Happily, at the start of the seventeenth century, Dutch opticians invented the telescope. By combining two lenses they found that they could magnify a distant object. Galileo applied the telescope to astronomy and made several major discoveries. He found the four large moons of Jupiter, and showed that the Milky Way is made of millions of faint stars.

A telescope's main job is to capture radiation from planets, stars and galaxies. This radiation may be in the form of light waves, radio signals or X-rays. Each radiation needs a special type of telescope.

With a big telescope an astronomer can collect far more radiation from faint objects than he could detect with his eyes alone. For example, the world's largest optical telescope, which is in the USSR, has a collecting mirror six meters (236 inches) across. When it looks at the stars, it has the seeing power of a million human eyes. Furthermore, a telescope can spend time collecting radiation from one object. It can build up a photograph, perhaps over many hours, of stars the eyes would never see no matter how long they stared into space.

A refracting telescope has a lens to gather light and to form an image of the object. This lens, the one at the front of the telescope, is called the objective. An eyepiece made from one or more small lenses is used to look at the image made by the objective. Strange as it may seem, astronomers are not always interested in using very high magnifications. Different eyepieces give different magnifications on the same telescope. Yet however much a star image is magnified it never looks like anything more than a fuzz of light!

The world's largest refracting telescope has an objective lens 1.1 meters (40 inches) across. Many problems make it impossible to manufacture lenses any larger than this. So for the biggest optical telescopes, astronomers instead use a curved mirror to reflect the light into an image.

*Inside the aluminium dome of the Anglo-Australian Telescope. The telescope is shown tipped over at a large angle to make it easy to see the enormous horseshoe mount. Behind the telescope is a control room. At night the astronomers operate computers that control the telescope and they can see what is happening by closed circuit television. Several similar telescopes are used in the Americas. All have reflecting mirrors about four meters (150 inches) across.*

Astronomers use reflecting telescopes for much of their work. In a large reflector, such as the Anglo-Australian Telescope, the astronomer can work inside a small cage positioned high up in the telescope tube. This cage may be replaced by a second curved mirror that reflects light back down the telescope and through a small hole in the main mirror. The light beam emerging from this hole is then examined by special instruments. The one most frequently used is the spectrograph. This breaks light down into its separate wavelengths. By examining the intensity of the light at different wavelengths astronomers can work out the temperature and composition of a star.

Radio telescopes use huge dishes as much as 100 meters (330 feet) across to reflect radio signals onto a radio detector located at their main focus. With a radio telescope, it is possible to measure the strength of the radio waves sent out by galaxies. By linking several radio telescopes together astronomers can also produce a "radio photograph" of radio-emitting regions of the sky. Radio astronomy can be done by day as well as at night.

High above the atmosphere there are X-ray and ultraviolet telescopes orbiting the Earth. To describe the view of the sky in X-rays and ultraviolet rays they send radio messages back to the Earth.

For amateur astronomers an excellent first instrument is a good pair of binoculars. With these an amateur may view details on the Moon, star clusters, and the moons of Jupiter. The next stage might be to obtain a small reflecting telescope with a 15 cm to 20 cm (six to eight-inch) mirror.

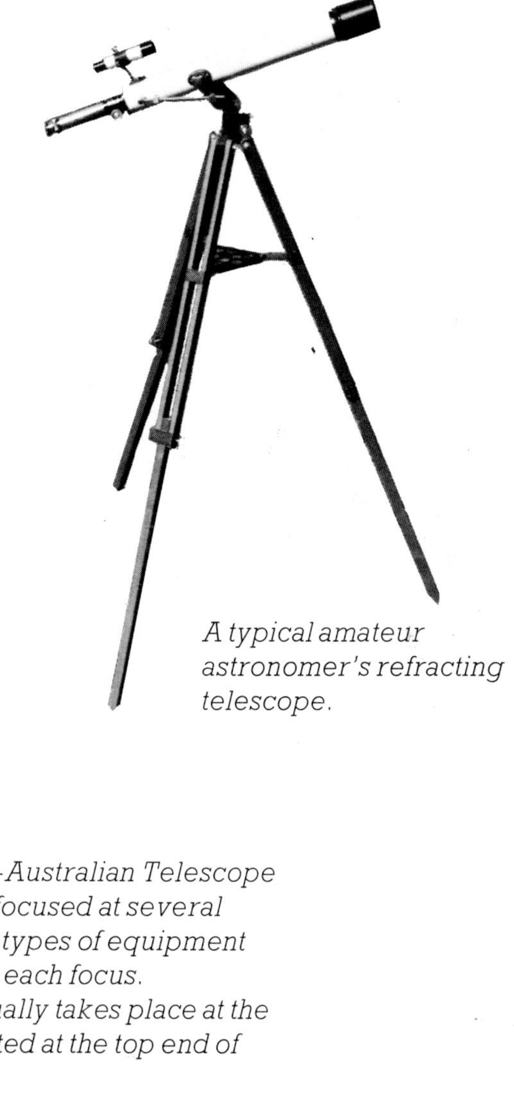

*A typical amateur astronomer's refracting telescope.*

prime focus

secondary mirror

Cassegrain focus

*Inside the Anglo-Australian Telescope the light may be focused at several places. Different types of equipment are positioned at each focus. Photography usually takes place at the prime focus located at the top end of the telescope.*

# Astronomy in Space

The space age has greatly changed astronomy. No longer do all observations have to be made through a restricting blanket of atmosphere. The atmosphere protects us by cutting out certain types of radiation that are constantly bombarding the Earth from space and would be dangerous to life. In particular, astronomers can only detect ultraviolet rays and X-rays if a telescope is placed in orbit round the Earth. These rays do not travel very far through air so they cannot reach an observatory at ground level. Another important advance made possible by space travel is the sending of television cameras to the nearby planets. Man has landed several times on the Moon. Venus and Mars have been probed by instruments that soft-landed on their surfaces.

The most spectacular achievement of space astronomy was the Apollo space program. This enabled men to make several Moon landings. For the first time they brought back material from another world for scientists to study. As a result, we now know that the age of the Moon is 4500 million years, the same as the Earth's.

It will be many years before travel even to the nearest planets. Mars, for example, would require an expedition lasting at least a year. Venus is far too hot and its atmosphere too dense for a visit, and Mercury is like a blast furnace. Instead, these planets have been explored by automatic cameras aboard space probes launched from Earth. In many ways, the technique of planetary photography is similar to a remote television broadcast. A television camera makes magnified images of the planet. These are transmitted back to Earth, line by line, as radio signals. Scientists can then view the planet on a TV screen.

Tens of thousands of photographs of Mercury, Venus and Mars have been made in this way. These have greatly increased our knowledge of the planets. Soviet astronomers obtained the first picture from the surface of a planet with their craft Venera 9 broadcasting from Venus.

*A large telescope in permanent orbit round the Earth can get beautiful views of stars and galaxies. This is because no clouds, rain or atmosphere disturb the seeing conditions. Space telescopes are controlled by radio signals from the ground. Information on the stars is also beamed back to astronomers by radio waves. If the telescope breaks down, it can be visited by engineers aboard the Space Shuttle. Astronomers hope that the space telescope shown here will be launched in 1985.*

American scientists subsequently probed the surface of Mars with the two Viking landers. Eventually, Viking landers will settle on satellites in the outer solar system, such as the moons of Jupiter.

X-ray telescopes have made surveys of the stars and galaxies that send out X-rays. Until instruments sensitive to these rays could be sent up to a height of 150 kilometers on rockets, astronomers knew nothing about X-ray stars. X-ray telescopes now circle the Earth permanently. They have detected many types of X-ray star. Some space scientists believe they have found X-rays coming from the vicinity of black holes in space.

An American orbiting laboratory named Skylab showed that astronomy would advance still further if optical telescopes were used in space. Then there is no interference from lighting, clouds or the atmosphere. A space telescope two meters in diameter would be able to see stars that are a hundred times fainter than any seen so far. It might even be able to find planets orbiting nearby stars.

*The spacecraft Voyager will take close-up pictures of the giant planet, Jupiter. This planet is 700 million kilometers from Earth so only a spacecraft can bring us really sharp images of it. In the late 1980s further probes will cruise past Jupiter and on to Saturn. Space probes also obtain many thousands of photographs of the planets of the inner solar system.*

*The United States Space Shuttle will be able to glide back to Earth after completing a mission. The Shuttle can carry astronomical telescopes for studying the radiations that do not get through the atmosphere of the Earth.*

# The Planet Earth

Man has been able to study the surface of his own planet for as long as the Earth has been inhabited. Yet, it is strange to think that before orbiting spacecraft had actually returned color photographs of Earth, nobody had predicted accurately what it would look like from space. Now we know the Earth as a beautiful blue and white planet. From beneath the spiralling patterns of brilliant white clouds the familiar shapes of the continents loom into view.

Many factors make the Earth unique in the solar system. It is the only planet with substantial amounts of liquid water. The oceans cover more than three-quarters of the surface; this vast quantity of water, coupled with the presence of oxygen in the atmosphere, is a powerful force of erosion. Shifting weather behavior and long-term changes in climate rapidly wear down the continental rocks. Mountains are smoothed by glaciers, wind and rain. Mighty rivers etch channels through the rocks and lowland plains, carrying sand from one place and laying it down in another.

Erosion has given the Earth a quite different appearance from that of the other planets in the inner solar system. The evidence that Earth was once pitted with meteorite craters in the same way as the Moon is now very scanty, but it is hard to imagine that Earth escaped this tremendous bombardment. Rather, the patient erosion by wind and water has healed over the wounds until no traces remain except in a handful of places.

*Blue oceans and white clouds dominate the view of Earth from space. Africa is visible beneath the clouds.*

*The crust of the Earth is like the skin of an apple. Beneath it lie the rocks of the mantle and the iron core.*

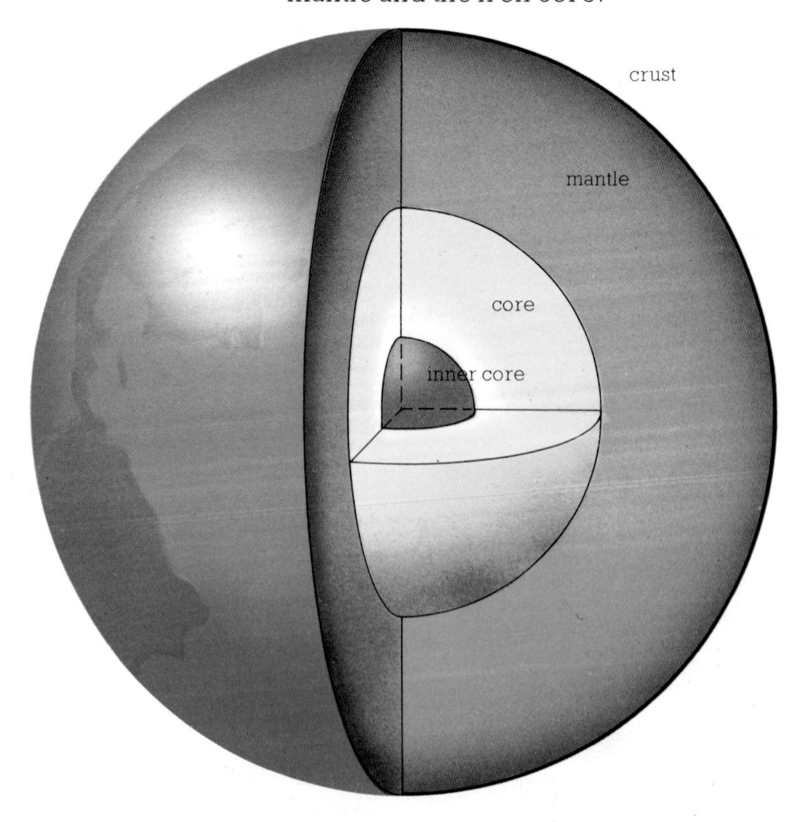

*Grand Canyon, Arizona, is a dramatic witness to the power of water. The Colorado river has now scarred the face of the desert right down to rock layers two thousand million years old. The bands of rock record a history of upheaval and flood, erosion and mountain building throughout that period of two thousand million years.*

Unlike the other rocky planets, the Earth has inner layers that are very active and on the move. Volcanoes and earthquakes are mechanisms that permit the Earth to let off pressure that builds up internally as the rocks beneath our feet slowly slide about. Earthquakes are sudden, unpredictable, and fatal in many parts of the world, but they teach geologists about the inner structure of the Earth. Vibrations spreading out from an earthquake are measured all over the globe. The manner in which these vibrations travel shows that Earth is made of several layers. We live on a very thin crust of solid rock that is nowhere more than 50 kilometers thick. The crust lies atop a thick layer of rock mineral called the mantle. Inside that, there is a liquid core of hot iron and nickel. Possibly the central part of this core is solid because of the immense pressure created by the weight of the overlying material. We know that the core is made of heavy substances such as iron because that's the only way that the Earth as a whole can end up with a density rather greater than that of ordinary rock.

Unlike many of the rocky planets in the solar system, our own world is a hive of geological activity. Mountains are being thrust up, earthquakes make the globe tremble, and volcanoes cough out liquid rock. The continents even are slowly gliding about. Why is Earth so different?

The answer is that the crust of the Earth consists of several large plates that will not keep still. Beneath the oceans and continents there is a rock layer that moves. Heat flowing from the interior of the Earth causes this motion, which is like that of a conveyor belt. The heat comes from the decay of radioactive rocks. In certain places the rock conveyor belts are pushing into each other, and tremendous buckling of the continental plates takes place. This crumpling of two continental plates has caused the formation of the Alps and Himalayas. Along the west coast of America the continental plate is being forced against the ocean plate and this has formed a great range of coastal mountains from Alaska to southern Chile.

Another effect of these rock movements is to generate friction. This may melt the rock just below the surface, which then bursts through a crack and makes a volcano. The Earth's largest volcanoes are the Hawaiian Islands.

The motion of continental and ocean plates is not noticeable in a human lifetime. But it is fast enough to change the face of the Earth in a mere 100 million years. For example, all the present continents resulted when two enormous land masses shattered about two hundred million years ago. South America and Africa are still drifting about but if you look at a map of the world it is easy to see that they once fitted together.

Perhaps you have used a magnetic compass to find directions. On most planets a magnetic compass would be of no use for finding north. The compass works here because the Earth has a magnetic field of its own. A compass needle lines up with the Earth magnetism and points to the north. Earth magnetism is generated by electric currents flowing through the liquid iron in the interior. Most planets lack this liquid metal so they do not have any magnetism.

Earth is not only our home, it is also the most active rocky object in the solar system.

# Mercury

Mercury is a world of extremes. As the closest planet to the Sun, it has scorching daytime temperatures of 400°C, hot enough to melt lead. At night the temperature plunges down, perhaps to –200°C because there is no blanket of atmosphere to trap the heat. Mercury spins round on its own axis in 59 days and takes 88 days to make one circuit of the Sun. Its orbit is quite elliptical, so that the distance from the Sun varies between 47 and 69 million kilometers. This tiny planet is not much bigger than our Moon.

In 1974, the space probe Mariner 10 flew past Mercury. It sent several thousand photographs of the planet back to Earth. At first sight, these show a world like the Moon. There are a great many craters and several smooth plains. Some of the large craters have bright rays spreading out just like the large lunar craters.

There is no weather on Mercury, just long baking days followed by intensely cold nights. The surface is, therefore, not changed by erosion. Over thousands of millions of years, space debris has smashed into the planet. This continuous bombardment has left the shattered surface we see today. The crater floors are covered in fine dusty powder, the product of smashed rock.

The heaviest bombardment of Mercury took place long ago, only a few hundred million years after the planet formed. Some of the material smashing into the surface scooped out craters hundreds of kilometers across. At the same time, volcanoes sprang into action, flooding craters with molten rock. Today, however, Mercury does not have any active volcanoes.

Mariner 10 detected a very weak magnetism round Mercury. It seems that the planet is like a permanent magnet. Earth's magnetism is a hundred times stronger than Mercury's.

An extremely thin atmosphere of helium gas was discovered by Mariner 10. There is so little of it that the surface pressure is billions of times smaller than at the surface of the Earth. Life like that on Earth could not exist on this airless world with its great range of temperature.

Mercury is a difficult planet to observe. Its close orbit round the Sun means that we never see it travel more than 27 degrees in the sky from the Sun's disk. Consequently, you can only see Mercury just before sunrise or after sunset, close to the horizon. Mercury's rapid progress limits the possibility of seeing it to just a few days per orbit. In a telescope it shows phases like the Moon, because from Earth we can see parts of the planet that are not in sunlight. If you search for Mercury with a telescope, be sure not to point the instrument at the Sun.

Earth

Titan

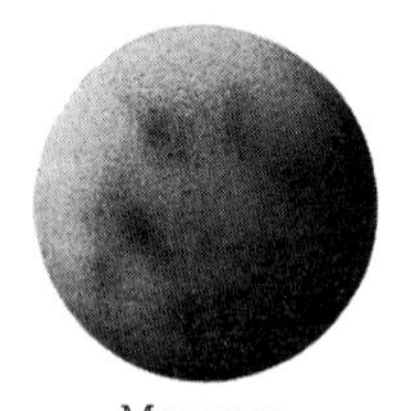

Mercury

Io

Moon

*Mercury is the smallest planet in the solar system. Here its size is compared to the planet Earth, the Moon, Io, a moon of Jupiter, and Titan, the giant moon of Saturn.*

water, no oceans, and no weather, so it has remained in much the same form for thousands of millions of years.

Astronauts left scientific apparatus on the Moon. Earthquake sensors detected numerous "moonquakes" as well as space rockets and meteorites slamming into the surface. Several small reflectors, like those on a car or cycle, were placed on the Moon. Scientists can now find the Moon's distance to within a centimeter or so by aiming a powerful laser beam at these reflectors and timing the round trip for laser light from Earth to Moon and back again. This distance on average is 390,000 kilometers. On the surface of the Moon you would weigh only one sixth of your Earth weight. This is because the Moon's mass is a mere one-eightieth of the Earth's, so the gravitational pull is considerably smaller.

The analysis of lunar material has shown that the surface has never supported life in the past. However, astronauts brought back to the Earth a piece of the Moon lander, Surveyor 3. Bacteria on this craft were still alive after several years of exposure to the harsh lunar environment. These bugs did not flourish, but neither did they die. Our spacecraft are contaminating the Moon, planets and deep space with the life from Earth even though the equipment is given a complete cleaning before its launch.

*An old crater, called Thomson, that has been flooded with lava.*

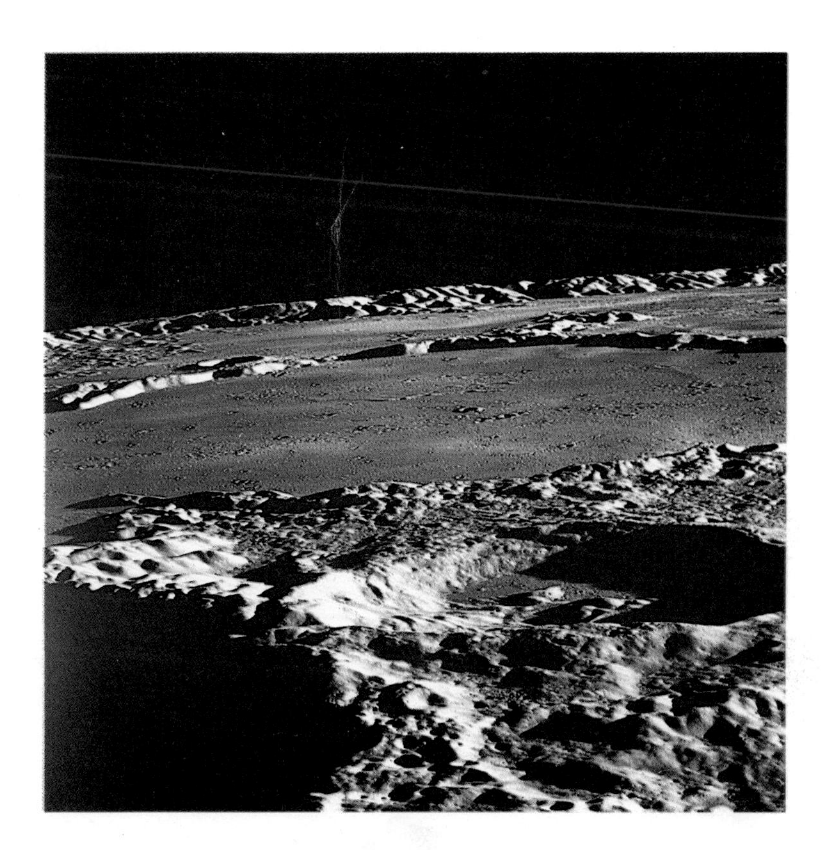

*Crater Tsiolkovsky on the far side, has a black carpet of lava and a central mountain.*

# Venus

Venus is one of the easiest planets to pick out in the sky, and is sometimes called the evening star. Brilliant Venus gets closer to Earth than any other planet: its near point is only 42 million kilometers away. At its brightest it outshines everything except the Moon and can be seen in the east at sunrise or the west at sunset.

Venus is the twin-planet of Earth because they both have the same size and mass. However, their atmospheres are very different. The main gases in our air are oxygen and nitrogen. Instead of these gases, Venus has a suffocating atmosphere of carbon dioxide. High in the Venusian atmosphere there are even misty clouds of sulphuric acid.

One remarkable property of the Venusian atmosphere is the way it acts on the planet like a greenhouse. The glass in a greenhouse lets high-energy heat rays from the Sun through to heat the soil. But the glass will not let through the low-energy heat that plants and soil produce. So heat is trapped inside the house and the temperature goes up. Now on Venus the dense carbon dioxide gas works rather like the glass in a greenhouse. It holds in the heat so the temperature on the surface is nearly 500°C, which is even hotter than on Mercury. There is so much carbon dioxide in the atmosphere that it crushes down with an enormous pressure nearly one hundred times higher than our atmosphere. On Venus the crush of the air is as great as that one kilometre beneath an ocean on Earth.

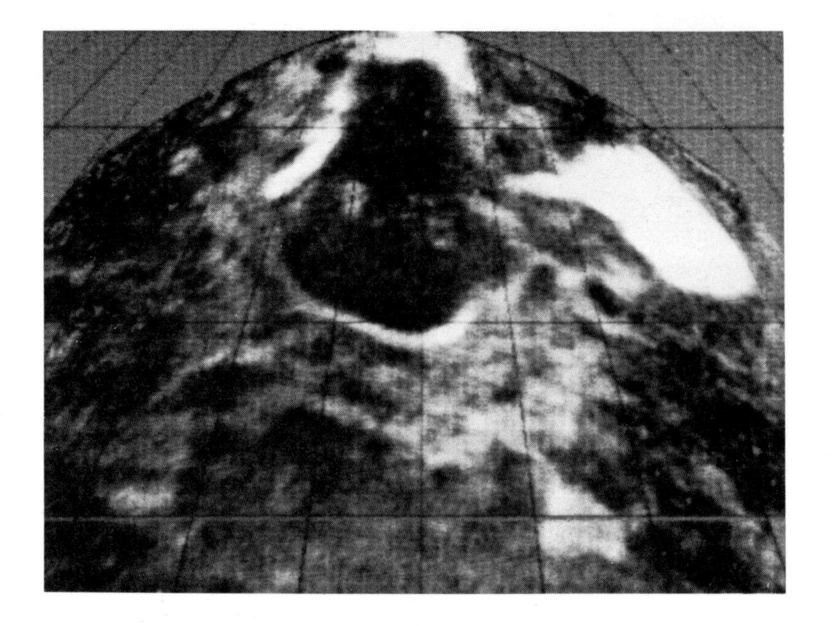

*Radar beams can penetrate the dense cloud. They reflect from the surface. Craters are faintly visible in this ghostly photograph made by radar beams.*

*The shattered rocks of Venus are visible in this distorted view of the surface. The Soviet spacecraft Venera 9 succeeded in sending back to Earth this panorama of a rock-covered surface. This was the first photograph taken on the surface of another planet.*

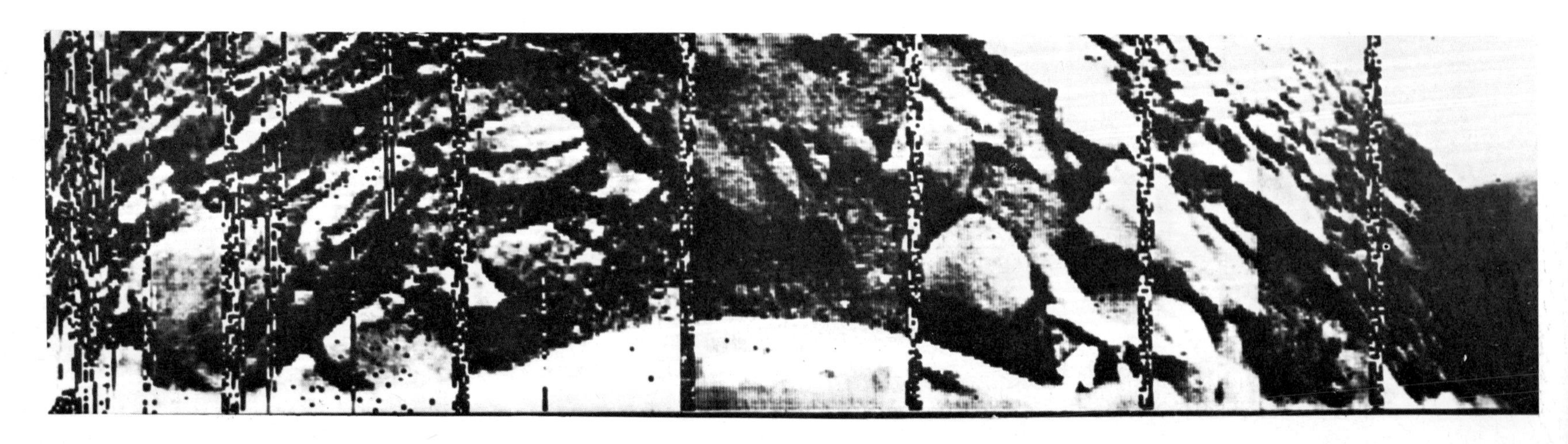

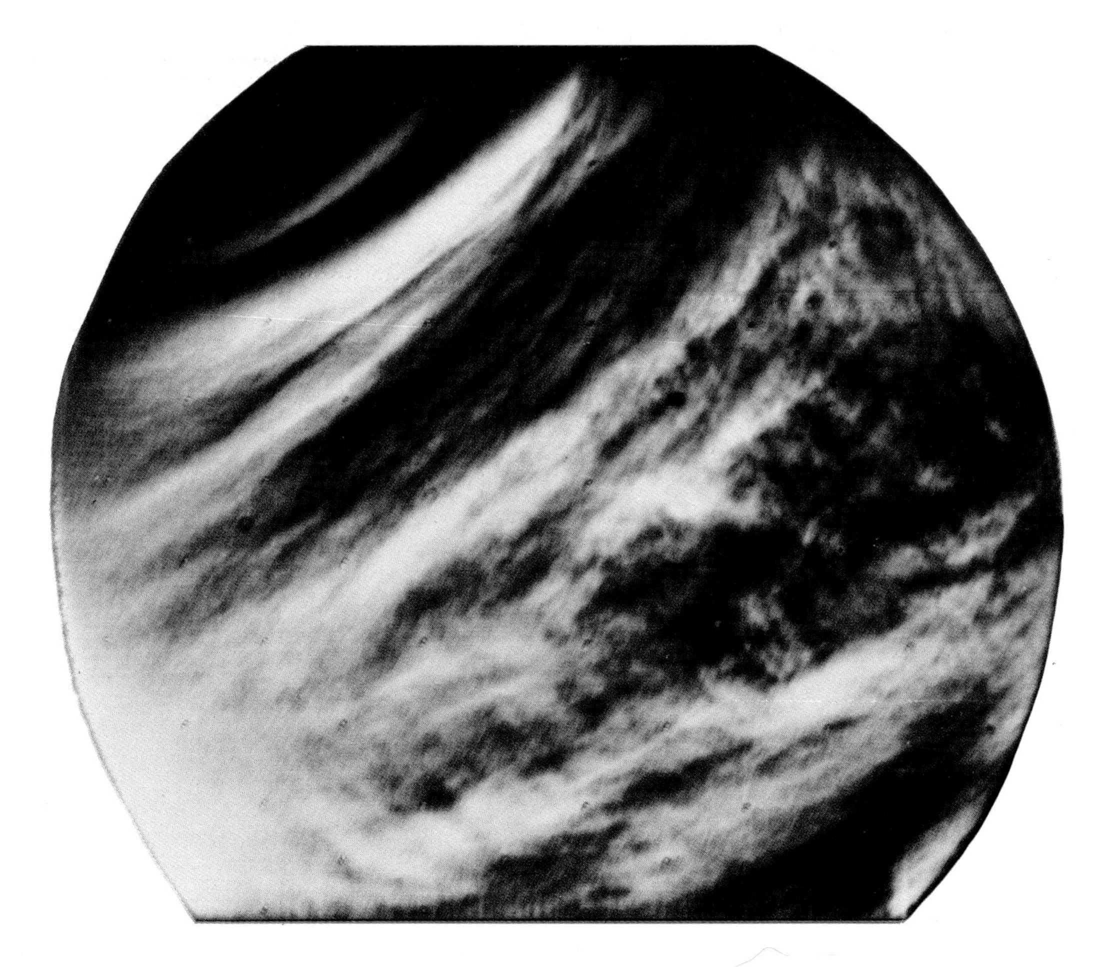

*Bands of clouds swirl high in the atmosphere of Venus. At the point where the Sun is shining directly overhead, the wind breaks the clouds into two streams, forming a Y-shape flow of clouds.*

Until the space age we didn't know very much about Venus. Swirling clouds hide the surface from view. By sending radar waves at Venus and listening to the echo from them, scientists have detected large craters and also mountain ranges about three kilometers high. Radar pulses also gave an amazing result for the rotation of Venus. This planet spins *backwards* on its own axis in 243 Earth days, whereas it takes 225 days to circuit the Sun. On Venus the solar day lasts for 118 Earth days; there are less than two Venus days in each Venus year!

The dense and hot world of Venus destroyed many spacecraft that tried to land on its surface. Finally, in 1975, two Soviet craft built like deep-sea diving probes survived for long enough to send us the first photographs ever taken on the surface of another planet. The pictures show a landscape of sharp rocks in one case and smoother features in the other case. Mostly the rocks are 30–60 centimeters in size.

The American probe, Mariner 10, took a look at the clouds of Venus as it coasted along on its journey to Mercury. Over three thousand photographs were taken. They mainly show the cloud formations in the upper atmosphere.

When Earth first formed, it may have been rather like Venus is today. Life on the Earth has broken down the dense carbon dioxide atmosphere which once existed here as well. Much of the carbon is now locked up in enormous layers of chalk and limestone. Water, which is still held in the atmosphere of Venus, deluged the Earth and formed the oceans.

If you observe Venus with even a small telescope, you will see that it has phases. Sometimes you can see a "half-Venus", and at other times a bright "crescent Venus". When Galileo discovered this behavior he realized that Venus must orbit the Sun at a closer distance than the Earth. This means that we on Earth sometimes see Venus from a vantage point allowing part of the nightside and part of the dayside to be seen together. When Galileo noticed this, most people still thought that the Earth lay at the centre of the Universe, with the Sun and Venus in orbit around it. The phases of Venus prove that this idea is wrong and that Earth and Venus both circle the Sun.

# Mars, the Red Planet

Viewed through a telescope, the planet Mars looks like a rusty-red disk. Its surface has various light and dark parts, as well as white ice caps at the north and south poles. Like our Earth, Mars experiences a cycle of seasons—while one half of the planet has summer, the other half has winter. In the summer half of the planet, the polar cap gets smaller as the ice melts in the Sun's warmth. Meanwhile, the other half of Mars is having winter, and the polar cap there grows.

During the course of a Martian year, which is nearly twice as long as an Earth year, changes occur in the appearance of the surface, especially near to the ice caps. The markings on Mars and their changes led astronomers to speculate for many years that Mars might have simple plant life. Earthbound telescopes could never answer questions about life on Mars for certain, but now Mars has been visited by spacecraft that have actually landed on it.

Viking I and Viking II both landed on Mars in 1976. They sent back to Earth many marvellous color photographs. The Viking craft also made experiments to find out about the soil and the atmosphere. Apart from these landers, several spacecraft have orbited round Mars. They have taken thousands of photographs from space so that a very great deal is now known about what Mars is really like.

The whole planet is a great desert. No water flows on the surface and practically none exists in the atmosphere either. Color photographs show a completely barren landscape, strewn with loose boulders. The red color is typical of desert rocks that are found in many places on Earth, and comes from the rusting or oxidation of iron. Only the surface is this color, for the Viking landers scooped up the red dust and found that just below the surface the rocks are a darker color.

Even the sky looks red on Mars, due to red dust in the air. Sometimes great dust storms develop, and about every ten years there is such a huge hurricane that the whole planet becomes engulfed in choking dust. Currents of air and very strong winds blowing at hundreds of kilometers an hour whip up the dusty surface into billowing clouds. When the storm dies and the dust eventually settles, it drifts to form lighter colored layers in some places. In other places, the dust is swept away to show the darker rock underneath. It is probably the winds that cause the changes in Mars' appearance by moving the dust around.

*These three photographs show changes in the appearance of Mars. The white polar ice cap is largest at the left (21 August 1971). The center view was taken on 9 October 1973, and only six days later (right photograph) the planet was partly obscured by a dust storm. These photographs were taken at the Lowell Observatory.*

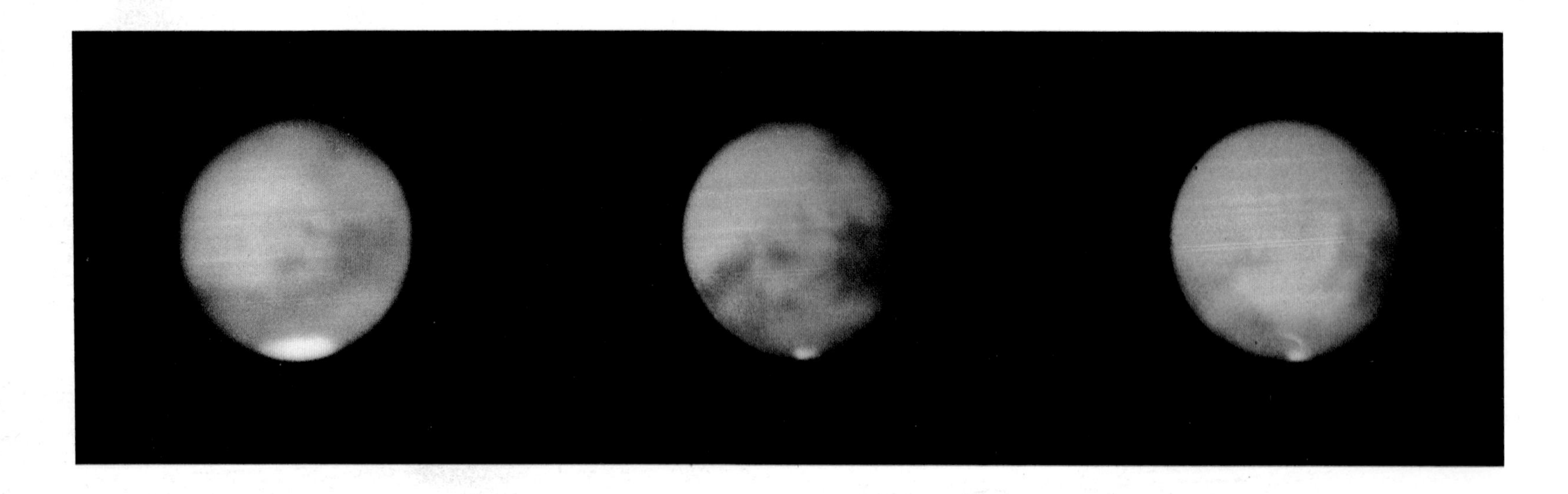

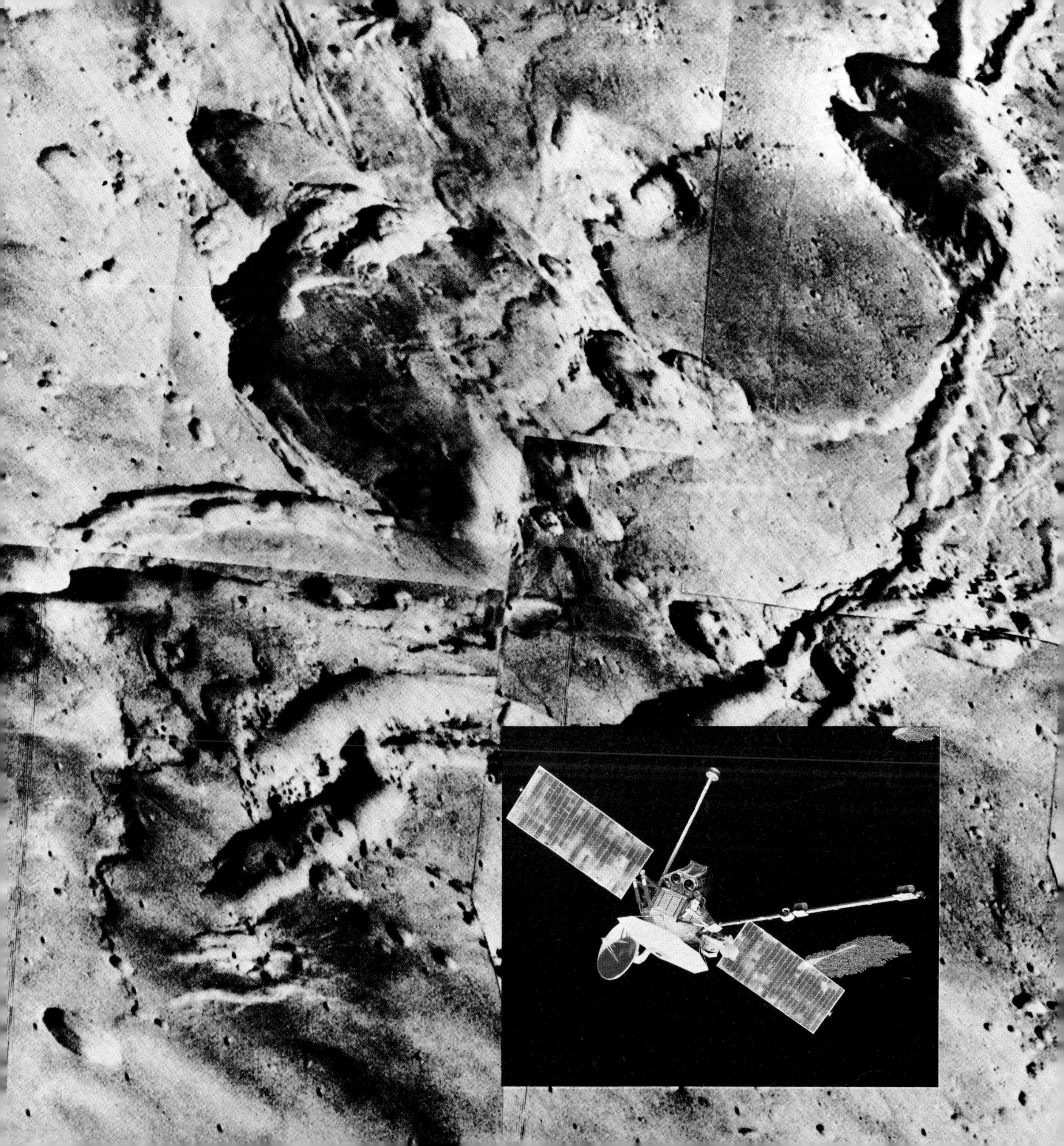

# The Surface of Mars

Like the Moon, Mars is pitted by many craters. These were created by meteorites that crashed onto the surface from space. Volcanic activity too has contributed to the scenery on Mars. In the region called the Tharsis Ridge, there are four, gigantic, extinct volcanoes. These mountains are called Olympus, Ascraeus, Pavonis and Arsia. Photographs taken from orbiting spacecraft show many old lava flows, long since cold and solid. The Viking landers took pictures of rocks that have a bubbly structure. This type of rock is made from volcanic lava. When the lava becomes solid, gas bubbles leave small holes in the rock.

Another dramatic feature is the Mariner Valley stretching nearly a third of the way round the planet. It makes the Grand Canyon in Arizona look very small, being three or four times as deep, and long enough to stretch halfway across the United States.

Mars is an inhospitable place. This cold, dry world has an atmosphere that is very thin indeed compared with our air. The Martian air consists mostly of carbon dioxide, so it could not be breathed by people or animals. There is almost no oxygen, and people could not live on Mars except inside special space stations.

We know now that there is no vegetation on Mars, not even simple plants such as moss and lichen. If there is any life at all, it must be in the form of a lowly plant life that has escaped detection. Most probably there is no life at all now, but it may have been very different in the distant past. Once, Mars had plenty of water flowing about in streams and rivers. We can see the old dried-up river beds and places where water has flooded over the surface. However, it must be a billion years or more since the last rain fell on Mars. In some places meteorites have made

*At the surface Mars is a red desert, devoid of life, with a scattering of sharp rocks. The sky has a pinkish color because of the dust that has blown into the atmosphere.*

*Mars photographed by the Viking spacecraft, showing several volcanoes.*

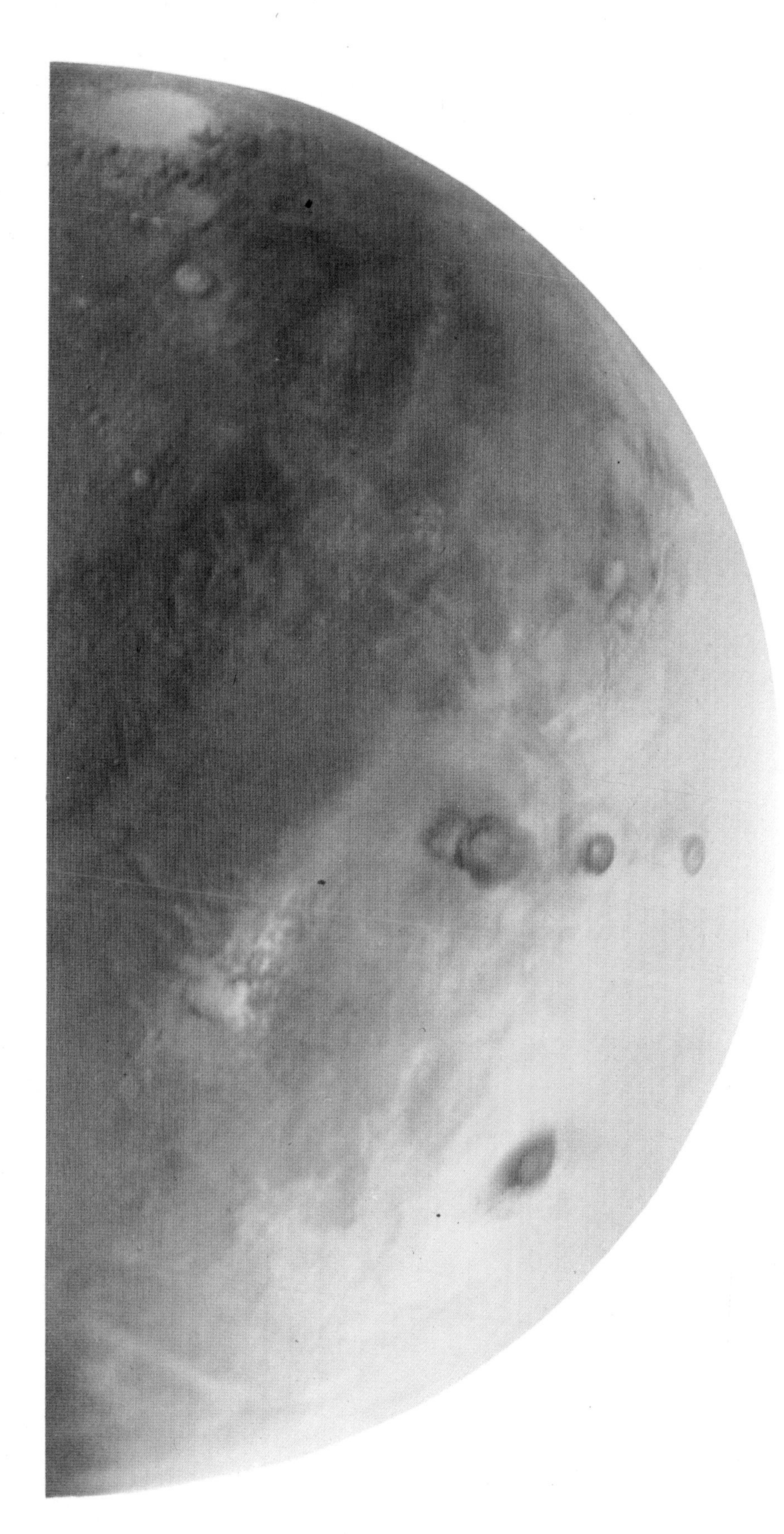

craters on top of the water channels. Small, thin clouds can be seen on many of the Mars orbiter photographs, but these cannot make any rain. Long ago the Martian atmosphere was much thicker, like Earth's. However, Mars is small, only just over half the size of the Earth. Its gravitational pull is less than Earth's and so the atmosphere has gradually drifted off into space.

Today the only water on Mars appears to be in the parts of the polar ice caps that never melt. The frost that comes and goes each year is actually frozen carbon dioxide, sometimes called dry ice. Close-up photographs of Mars' polar caps show a surprising effect. All round the pole, dark bands curve in spiralling patterns. These bands are valleys that are free of frost. Inside the valleys are rocky terraces revealing many layers of mixed snow and dust. The winds of Mars circulating around the polar caps have caused the valleys to be eroded away, and ridges left standing. Round the north polar region there is a giant ring of sand dunes. The red sands are constantly shifting on this dry, desert world.

*Part of the Mariner Valley, a vast chasm on Mars that is in places 40 kilometers wide and several kilometers deep. Rivers, now dry, have cut deep channels in the side of this valley.*

# Jupiter and Saturn: 1

Jupiter, Saturn, Uranus and Neptune are giant planets in the outer solar system. They have the major share of all the planet matter. The giant planets have over a hundred times as much material as the tiny planets circling the inner solar system. The outer giants are mainly made of light gases, such as hydrogen and helium, whereas the inner planets are made of rocks and iron.

The solar system giants are considerably larger than the inner planets. Jupiter, for example, is eleven times the diameter of the Earth and it has a volume over one thousand times as large. These giants are not so dense as the Earth either, for their densities are closer to that of water than of rock. Saturn, in fact, is not even as dense as water—a lump of Saturn matter would float on the sea.

All of the giants spin rapidly on their rotation axes, Jupiter taking less than ten hours to make a single turn. This high-speed twirling makes the planets bulge out at their equators. A further interesting feature of the outer solar system are the many moons, over thirty in all. Jupiter and Saturn each have a moon that is slightly bigger than planet Mercury. And Saturn, of course, has intrigued us for centuries with its splendid system of rings.

Jupiter, the nearest of the giants to the Sun, is also the largest and most massive. At times it outshines all the stars in the night sky—only planet Venus gets brighter. Jupiter shows a variety of features, some of which you can see in a small telescope. The dark and light colored bands of cloud are prominent. You can see the Great Red Spot as well. This varies in size, but is nearly 50,000 kilometers long and 12,000 kilometers wide.

When we look at Jupiter we can only see the clouds and storms in the upper atmosphere. Even telescopes on spacecraft cannot glimpse a surface beneath the thousands of kilometres of foggy gas. The same is true of Saturn, Uranus and Neptune: only the clouds can be seen. Jupiter's clouds are made of a deadly mixture of methane (natural gas is methane), ammonia, helium, and water. The highest layer that we can see is made of frozen ammonia. Traces of other gases color the dark cloud belts.

In Jupiter's atmosphere there are high winds. Near the equator the clouds dash along at 360 kilometres an hour. The Great Red Spot is the most fantastic hurricane in the solar system which will keep blowing and twisting round for many thousands of years. Photographs taken by the American Pioneer spacecraft show the whirlwind inside the Red Spot.

If Jupiter is carefully observed through a telescope, changes are apparent in only an hour or so. This is because the planet spins fast, and in less than one night the entire planet may be seen. Besides watching the changes in Jupiter's weather it is also fascinating to follow Jupiter's four large moons.

In the year 1610 Galileo discovered Jupiter's four main moons. They are named Io, Europa, Ganymede, and Callisto. All four are easily spotted even with binoculars. Io takes just under two days to orbit Jupiter; Europa requires three and a half days, Ganymede a week, and Callisto nearly seventeen days. If a sketch is made of Jupiter and its satellites for several nights, the dance of the moons around the planet is soon revealed. Galileo realized that Jupiter is a miniature solar system. In addition to these four large moons, which are called the Galilean Satellites, Jupiter has at least nine much smaller satellites. They can only be photographed by large telescopes.

*As Saturn orbits the Sun, the tilt of the rings slowly changes. This is because we on the Earth see the rings from a steadily changing angle as Saturn moves along its orbit.*

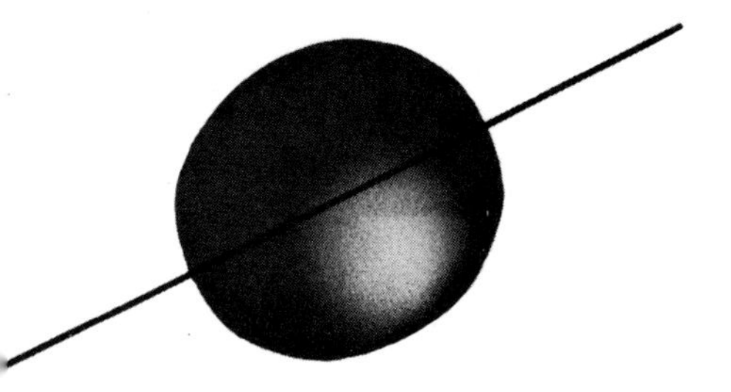

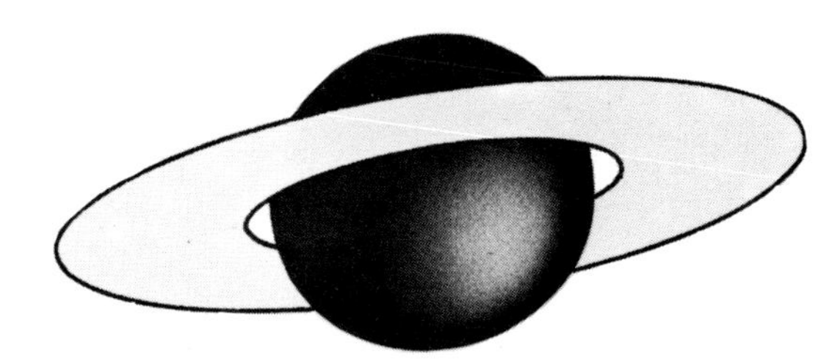

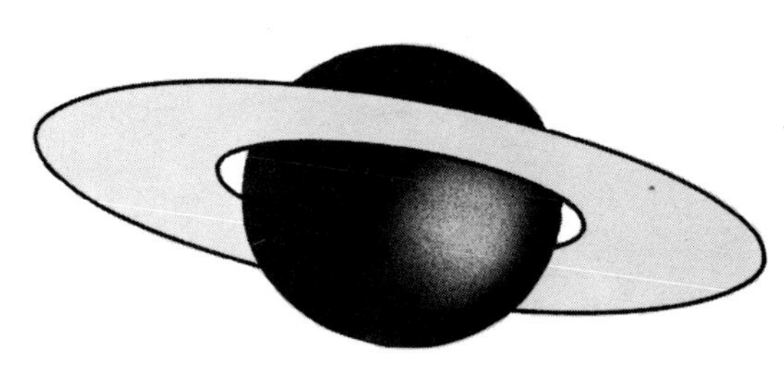

It's fun to watch the moons of Jupiter disappear behind the planet and then pop out again into the sunlight. Sharp-sighted observers have noticed that when Io emerges from the icy shadow behind Jupiter it looks brighter than usual for a few minutes. This is because ammonia "snow" drifts down from Io's atmosphere as it shivers on the night side of Jupiter. This brilliant snow reflects sunlight more strongly for a few moments before melting away in the Sun's rays. Io is coated in the yellow metal called sodium. This evaporates in the Sun's rays to form a deadly sodium atmosphere.

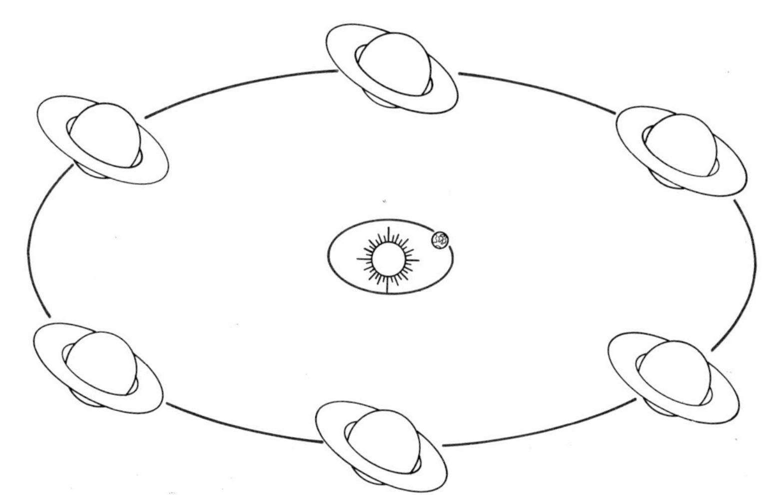

*Twice in each 29-year orbit the plane of the rings of Saturn cuts through the view line of Earth. At such times the rings are "edge-on" for a few weeks and they may seem to vanish.*

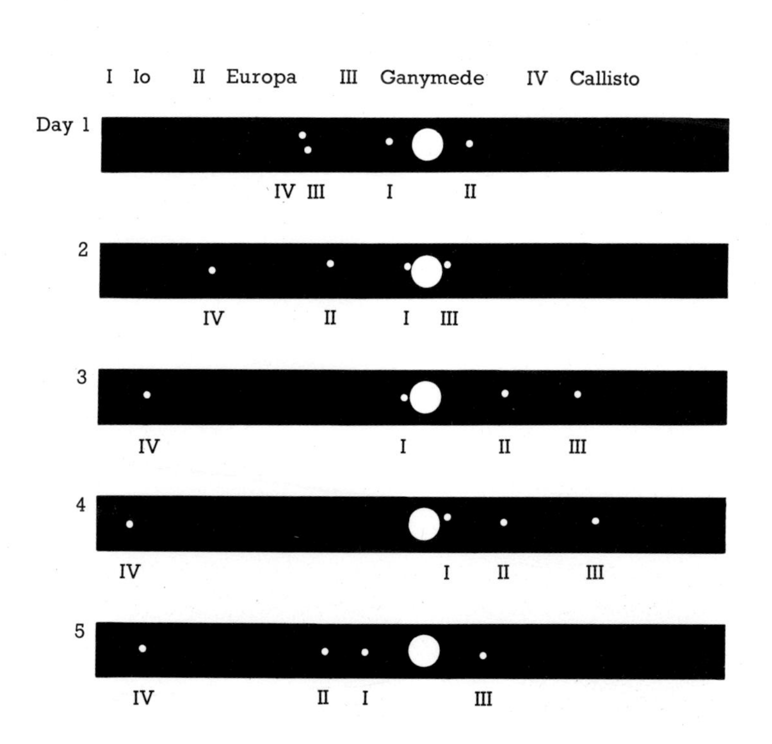


*The four large moons of Jupiter can be followed from night to night with a small telescope. On each night the arrangement is different. By sketching roughly the positions of these moons on several successive nights you can soon learn to identify each one individually. Io changes position the fastest because it has only a two-day orbit. Callisto is the slowest. The positions over a typical five-day period are sketched here. Sometimes one or more moons is out of sight, round the back of the planet.*

# Jupiter and Saturn: 2

Saturn is similar to Jupiter in many ways, except that it has a magnificent series of rings. Its mass is nearly one hundred times greater than that of the Earth, but almost all of this consists of gas and not rock. The cloudy belts are not so obvious as they are on Jupiter, but they certainly exist. Saturn has ten satellites. One of them, Titan, is nearly 6000 kilometers across, which makes it the largest moon in the solar system. Titan also has its own atmosphere, made of methane and ammonia.

Saturn has four rings in all. The innermost one is very faint and it almost touches the cloud tops. The bright outer ring is almost 140,000 kilometers in diameter. The gap between this ring and the next is caused by the gravity pull of the giant moon, Titan. This gap is called the Cassini Division.

The rings lie at an angle to Saturn's orbit round the Sun. One effect of this is to vary the openness of the rings as seen from the Earth. Twice during Saturn's 29-year journey round the Sun we see the rings open to the fullest extent. Halfway between these wide-open positions we see the rings edge-on, and, except in powerful telescopes, they seem to vanish. This shows that the rings are only five kilometres thick. Distant stars shine right through them with hardly any dimming.

Saturn's rings are made of myriads of tiny particles mostly only a few centimeters across. Each particle orbits the planet as if it were a tiny moonlet. By bouncing radar signals off these rings scientists have found that there are a few large boulders present as well as swarms of pebbles. The rings may have formed at the same time as the ten satellites. It is also possible that one moon strayed too close to the planet and got torn into tiny pieces.

*Saturn's ring system is magnificent when tilted wide open. The shadow of the planet falls on the far side of the rings. Saturn's cloud belts are not so easy to see in a small telescope as those of Jupiter.*

The giant planets contain a great deal of hydrogen. Far inside Jupiter there is a very strange world indeed. About halfway to the center the crushing weight of the planet causes the hydrogen to alter completely. Instead of being a gas it turns into a solid metal.

Jupiter has a very large magnetic field stretching far into space. Charged particles emit radio signals when they fly through this magnetism. Radio telescopes can map out the magnetic cage surrounding Jupiter by picking up these radio signals.

Another odd fact about Jupiter is that it sends out more heat than it receives from the Sun! This is because Jupiter is still shrinking, by about one centimeter each year. This shrinking releases heat energy. Jupiter would not need to be much more massive than it already is to turn itself into a real star, shining by its own light.

*Pioneer 10, an American probe launched into deep space, obtained this stunning view of Jupiter. At lower left is the Great Red Spot. The dark disk is the shadow of Io cast by sunlight.*

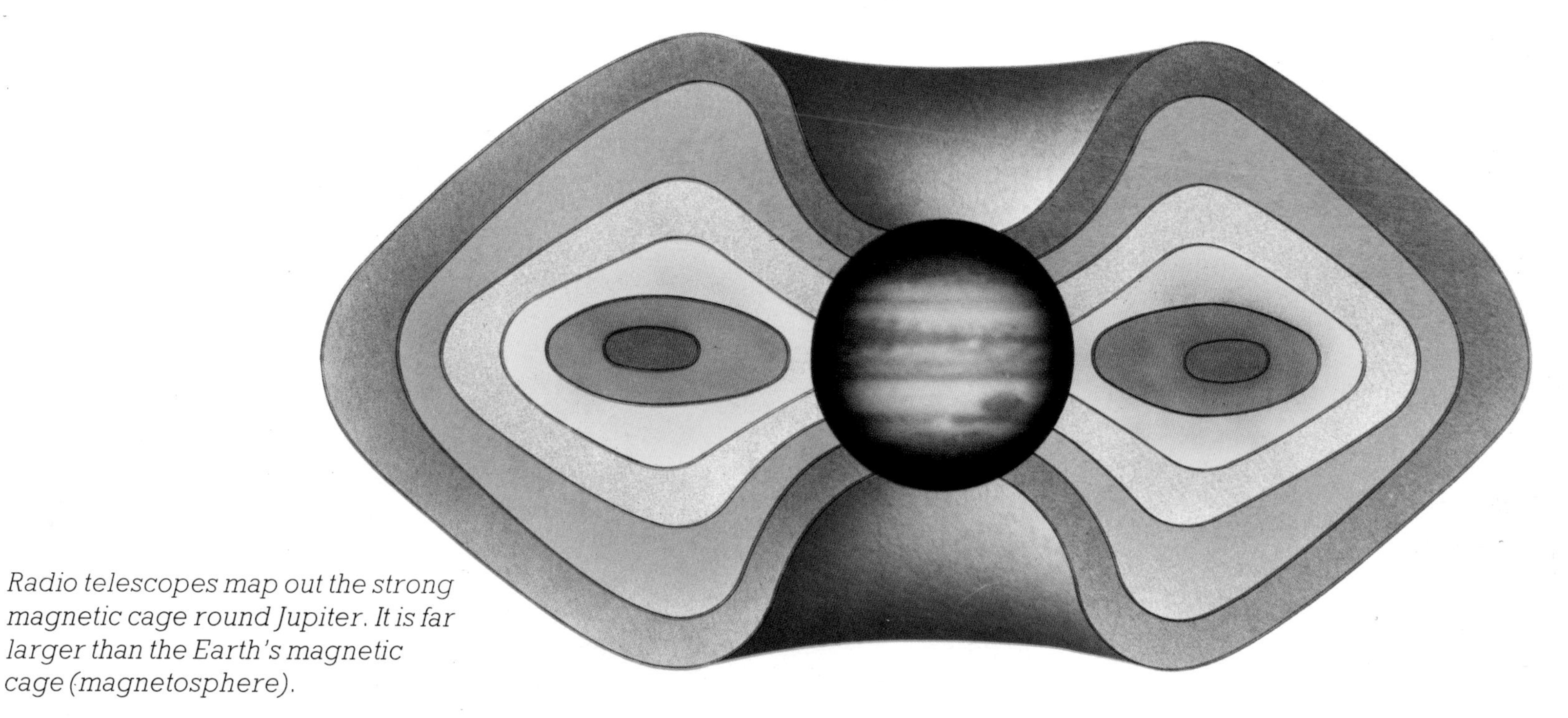

*Radio telescopes map out the strong magnetic cage round Jupiter. It is far larger than the Earth's magnetic cage (magnetosphere).*

# The Outer Solar System

Beyond Saturn lie Uranus, Neptune and Pluto, three planets unknown in ancient times. The great astronomer William Herschel found Uranus on 13 March 1781. Although Herschel had intended to work as a musician he found astronomy fascinating. He taught himself about the skies, and then in 1773 he made his own reflecting telescope. With this he started to look at the stars. As his enthusiasm and knowledge grew, he built larger and larger telescopes. His sister Caroline helped him to search the skies.

Herschel decided to make a survey of the stars and to note down their positions and brightnesses. During one of these careful searches he found an entirely new planet. This greatly surprised scientists who had not suspected that there were any further planets. Herschel wished to name the new planet after King George III who then reigned in England. Eventually, however, it was agreed to call the planet Uranus. This choice was made because in mythology Uranus is the father of Saturn, and Saturn the father of Jupiter. Herschel made many important observations of stars, galaxies, and nebulae in addition to finding the planet Uranus.

Astronomers located Neptune in 1846 after a remarkable piece of detective work by mathematicians. After many years of careful observation, Uranus puzzled observers. It did not keep to the path around the Sun that astronomers thought it should on the basis of Newton's law of gravity. Something kept knocking it off course! A young Cambridge mathematician named John Adams and a Frenchman, Urbain Le Verrier, both realized that another planet might be tugging Uranus to one side. These wizards computed where the unseen planet must be. Adams had no luck in persuading the Cambridge professor of astronomy to make a quick search. Instead, an observatory in Berlin worked on Le Verrier's calculations. In 1846 they found a new dot of light in the constellation Aquarius: planet Neptune had been found.

Soon Neptune started to go off course as well. Could there be yet another planet further out that was pulling Neptune to one side, astronomers wondered? In 1915 Percival Lowell worked out where it must be but nothing could be seen. Then, in 1930, Clyde Tombaugh found the ninth planet, almost by accident, after a long search. He named it Pluto.

Every so often, newspapers report that someone has predicted the existence of a tenth planet out beyond Pluto and even calculated where it must be. All these sensational claims eventually turn out to be based on wrong calculations. Planetary scientists now believe that there cannot possibly be a large unknown planet in our solar system. If it really existed we should by now be well aware of its gravity pull on the other planets.

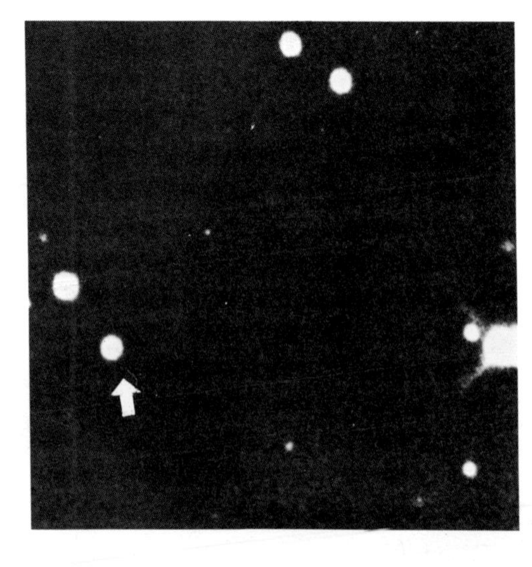

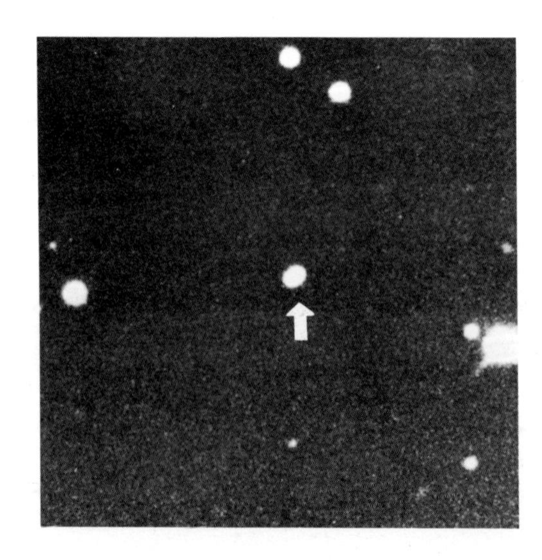

*Pluto is so distant that it looks like a star on a photograph. But it is truly a planet: it moves gradually against the distant starry background, as shown in this pair of photographs taken a few days apart.*

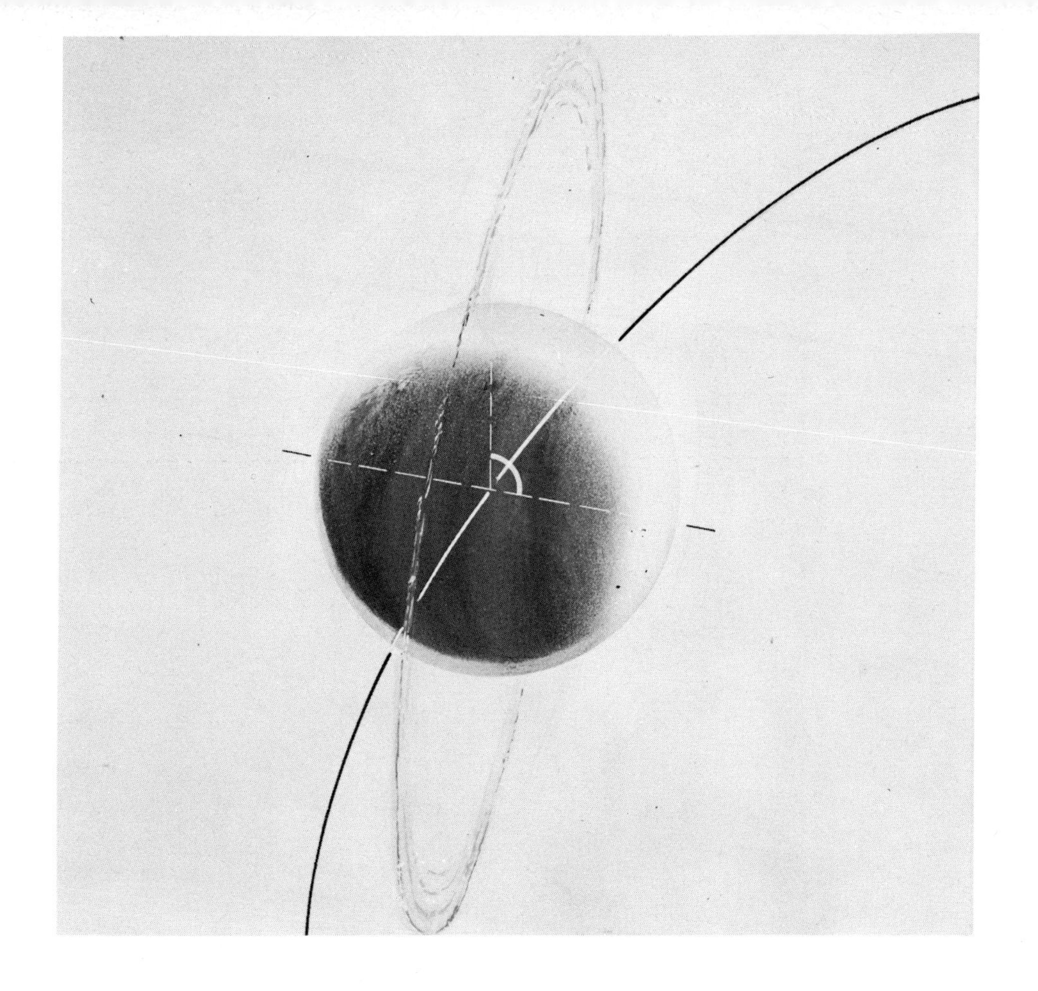

*Uranus probably has five necklaces of boulders circling round it. This ring system is very much fainter than Saturn's and it cannot be seen in even large telescopes.*

*Five moons orbit Uranus.*

Of course, there may be countless millions of tiny objects a long way off, all too far away to show in telescopes. For example, in 1977 a rock several hundred kilometres across was discovered to be drifting along beyond the orbit of Saturn. But this is a new type of asteroid rather than a genuine tenth planet. Astronomers have decided to name this far-flung asteroid Chiron.

Uranus has a diameter four times larger than Earth and it takes 84 years to orbit the Sun. Occasionally you can just about catch a glimpse of Uranus with the naked eye on a very dark night if you know exactly where to look. With an up-to-date chart of planet positions you can find it with a small telescope.

One odd feature about Uranus is that it is a planet lying on its side. The rotational axis is tipped over at an angle of 98°. This means that the seasons on Uranus must be very strange indeed. For several Earth years the Sun does not shine at all in one hemisphere, while the other is weakly bathed in sunlight.

In 1977 Uranus passed across the sightline to a distant star. All round the world, and even in a high-flying aircraft, astronomers trained their telescopes on Uranus to watch the planet eclipse the star. But they got a great surprise, for the starlight was blotted out several times before the planet actually cut across the line of sight. What had happened was this: Uranus must have a system of five rings that are invisible to our telescopes. Perhaps they are made of large boulders rather than sand and grit. As the planet and its rings tracked along against the distant background of stars, the ring-boulders blacked out the star for a fraction of a second. In addition to this ring system, Uranus also has five moons.

Pluto is a small world about 6000 million kilometers from the Sun. The orbit is an ellipse that crosses Neptune's path. By 1989 Pluto will actually be nearer to the Sun than Neptune and will continue to be so for some years. After this, it swings out again and will reach the greatest distance in about the year 2100.

We don't know very much at all about Pluto. It is probably as dense as the Earth but has only one-tenth the mass. There is no cloudy atmosphere—it's much too cold for that! Pluto has a freezing cold coating of ice, which is dimly lit by a bright *star*—the distant Sun. Possibly this icy outpost of the solar system was once a moon of Neptune that somehow got flung out of orbit and left to wander round the Sun on its own.

# The Asteroid Belt

After Herschel found Uranus, astronomers all over Europe swept the skies to search for more planets. The discovery of another planet would bring instant fame, and this spurred on the quest. Furthermore, a mysterious property of planet orbits, known as Bode's Law, seemed to show that somewhere a planet was missing.

In 1772 Johann Bode publicized a number game first discovered some years earlier. The number game works like this: start by writing 0 and 3, and then add to the string of numbers by doubling the previous number. This gives the list of numbers 0, 3, 6, 12, 24, 48, 96, 192, and 384. Now add 4 to each, making instead the series 4, 7, 10, 16, 28, 52, 100, 196, and 388. The table now shows a truly remarkable fact: if we take the number 10 to represent the Earth's distance from the Sun, then the number game correctly predicts most of the other distances. But in Bode's time there was a gap between Mars, at 16, and Jupiter at 52. This vacancy at number 28 seemed convincing evidence that an unknown planet lurked somewhere between Mars and Jupiter.

On 1 January 1801 G. Piazzi discovered a new object between Mars and Jupiter. He was actually making a list of stars when he spotted an extra object. Further observations showed that it was moving along an orbit between Mars and Jupiter, and that this orbit fitted well to Bode's number game. The object was named Ceres, but it is only 400 kilometers across and thus does not qualify as a real planet. So the search continued and, instead of locating a true planet, it revealed a swarm of minor planets, or asteroids, cruising round the Sun between the orbits of Mars and Jupiter. Today there are many thousands to keep track of and new ones are being found all the time.

Nearly all the asteroids, or minor planets, are between Mars and Jupiter. A small number, however, come closer to the Sun. In 1937, asteroid Hermes passed within 780,000 kilometers of Earth, barely double the distance of the Moon. If an asteroid ever succeeded in causing a direct hit it would be unimaginably destructive. Perhaps asteroid crashes caused the largest craters on the Moon.

Most of the asteroids are irregular lumps of rock measuring a few kilometers from end to end. They have rocky outcrops and craters but no soil or dust. Some are as black as soot and others a reddish color.

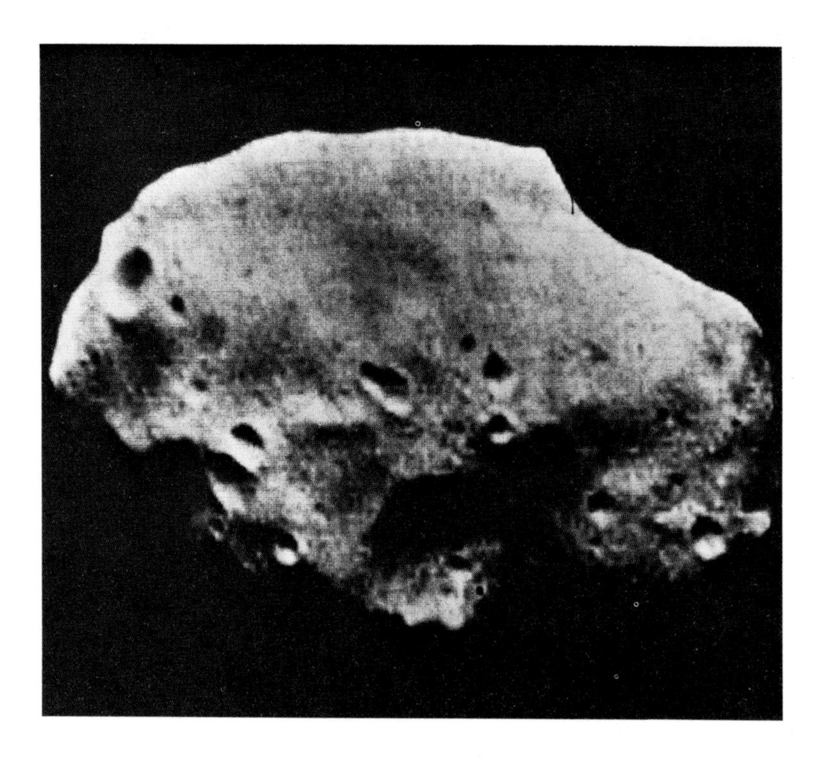

*Phobos, a moon of Mars, is very similar, perhaps, to a small asteroid. It has many craters, the scars of past encounters with meteoroids.*

**Johann Bode's planet number game**

| Planet | Bode's series of numbers | Actual distance from Sun with distance as 10 units |
|---|---|---|
| Mercury | 4 | 4 |
| Venus | 7 | 7 |
| Earth | 10 | 10 |
| Mars | 16 | 15 |
| (Asteroids) | 28 | (25–40) |
| Jupiter | 52 | 52 |
| Saturn | 100 | 95 |
| Uranus | 196 | 192 |
| Neptune | — | 300 |
| Pluto | 388 | 394 |

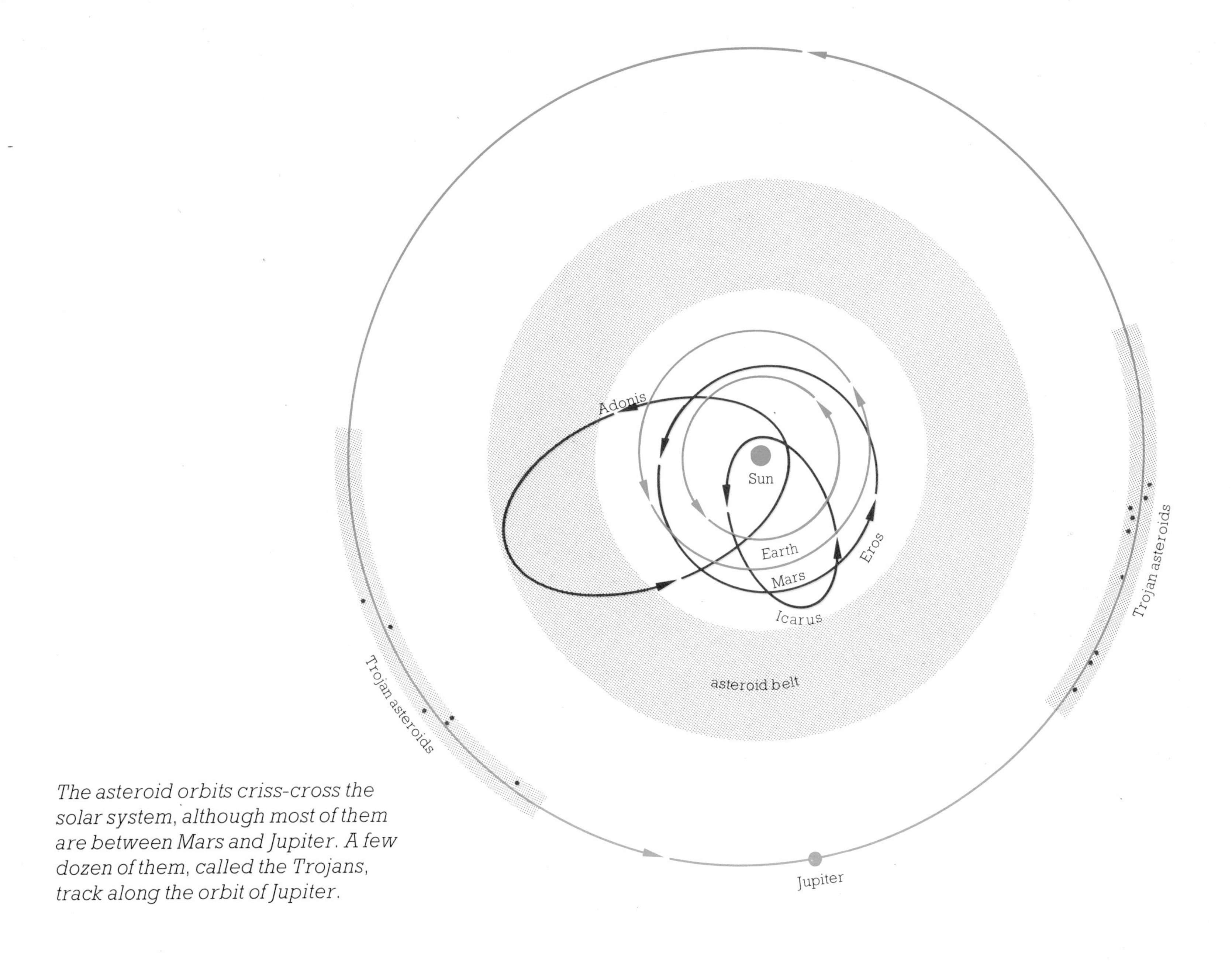


*The asteroid orbits criss-cross the solar system, although most of them are between Mars and Jupiter. A few dozen of them, called the Trojans, track along the orbit of Jupiter.*

At one time, people believed the asteroids to be the wreckage of a planet. However, the total amount of material in the asteroid belt is less than one-hundredth the Earth's mass. In all probability, the asteroids are the remains of material that got left over when the nine planets formed about five thousand million years ago.

Some idea of what an asteroid looks like is given by the photographs of Phobos and Deimos, the two midget moons of Mars. Close-up cameras have revealed details only a few kilometers across. Each mini-moon is sprinkled with small craters and sharp ridges. If you stood on an asteroid of this size—say 20 kilometers in diameter—the slightest movement would send you drifting into space because its gravity pull is very small.

In the far future, people may try to colonize the asteroid belt by constructing giant space stations on the large asteroids. The space stations would be totally self-supporting and thousands of people would live in each one. Rocket motors would deflect the asteroid colonies into warmer parts of the solar system. The colonists would sell rare minerals to the people on Earth and beam solar power down to them in the form of radio waves.

# Comets

About every ten years or so, you might see a bright comet trail across the sky. A good bright comet will remain visible for several weeks. Astronomers know of hundreds of comets. About two dozen of these visit our part of the solar system each year. Hardly any of them get bright enough to view without a telescope. The rareties that do blaze forth are among the most splendid sights in the heavens.

When a comet first comes along it looks like a fuzzy patch of light. Nightly it grows brighter as it heads along its path towards the Sun. Many comets grow a shimmering tail, which is so transparent that stars shine right through it. Sometimes the tail splits into two or more parts.

In the past people often feared that a comet would bring bad luck. Sometimes they were right! Halley's Comet appeared in 1066, at the time of the Norman invasion of southern England. In 1812, Napoleon's mighty army had to retreat from Russia, soon after the great comet of 1811 blazed through the skies. In fact these were just coincidences; comets are journeying harmlessly round the Sun and they have no influence whatsoever on Earth.

Records of comets go back for thousands of years. Some comets are regular visitors. Halley's Comet has been coming round every 76 years or so since before the birth of Christ. We shall next see it in 1986. Unfortunately, the viewing conditions for Europe and North America are not expected to be good, but in Australia and New Zealand they should be excellent.

New comets come from the furthest parts of the solar system. Most of them swing past the Sun in just a few months and then head back out again, finishing up far beyond Pluto. Each circuit of their highly-stretched orbits takes many thousands of years. Unlike planets, comets can be switched into entirely new orbits. They are flimsy objects and the gravity tug of a large planet, such as Jupiter, throws them into a new orbit if they stray too close. This is what happened to Halley's Comet, so that it now comes along more often.

As a comet draws close to the Sun, its tail grows. The tail always points away from the Sun. The pressure of sunlight and the rush of wind from the Sun push the tail away. Every time a comet passes the Sun, it loses material. Encke's Comet, which has made dozens of passages since discovery, is now almost exhausted. Soon it will be completely defunct.

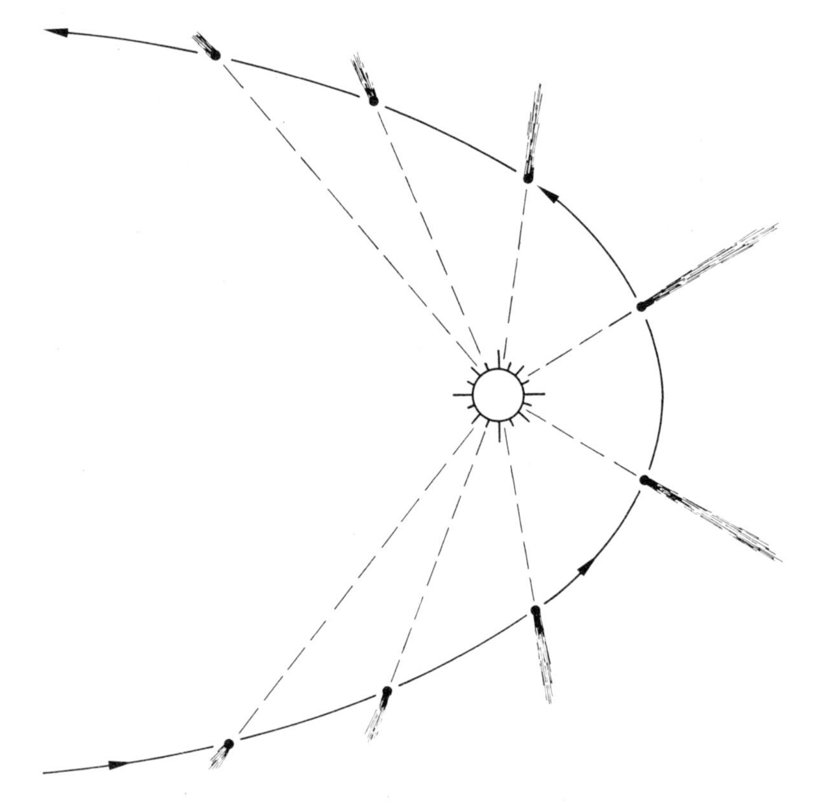

*As a comet moves on its orbit round the Sun, the tail changes in size and direction. It always points away from the Sun, no matter which direction the comet travels in, on account of the pressure of sunlight and the solar wind, both of which push the tail away from the Sun. As the comet gets closer to the Sun, heating effects cause the ejection of gas and dust to increase. In this way the tail grows in size. Then, after passing the closest point of the orbit to the Sun, the action starts to decline and the tail shrinks.*

A comet is probably made of gas, ice and gravel all frozen into a ball. When it gets nearer to the Sun, it warms up. Gas and dust then stream off. Finally the comet head freezes up again as it travels away from the Sun.

Comets are normally named after their discoverers. Many of the new ones are found by amateur astronomers who specialize in comet searching. Several people may see the same new one in only a few hours, and then the comet may have up to three names attached to it. There have been a few exceptions to this rule for naming comets. Edmond Halley computed the orbit of the comet that now bears his name, and comets found by Chinese scientists are named after their observatory.

*Comet West trailed across the sky in 1975, displaying a beautiful fan-shaped tail as it did so. This comet could be seen with the naked eye. Brilliant naked-eye comets are rare sights indeed, and a great many comets can be viewed only with powerful telescopes.*

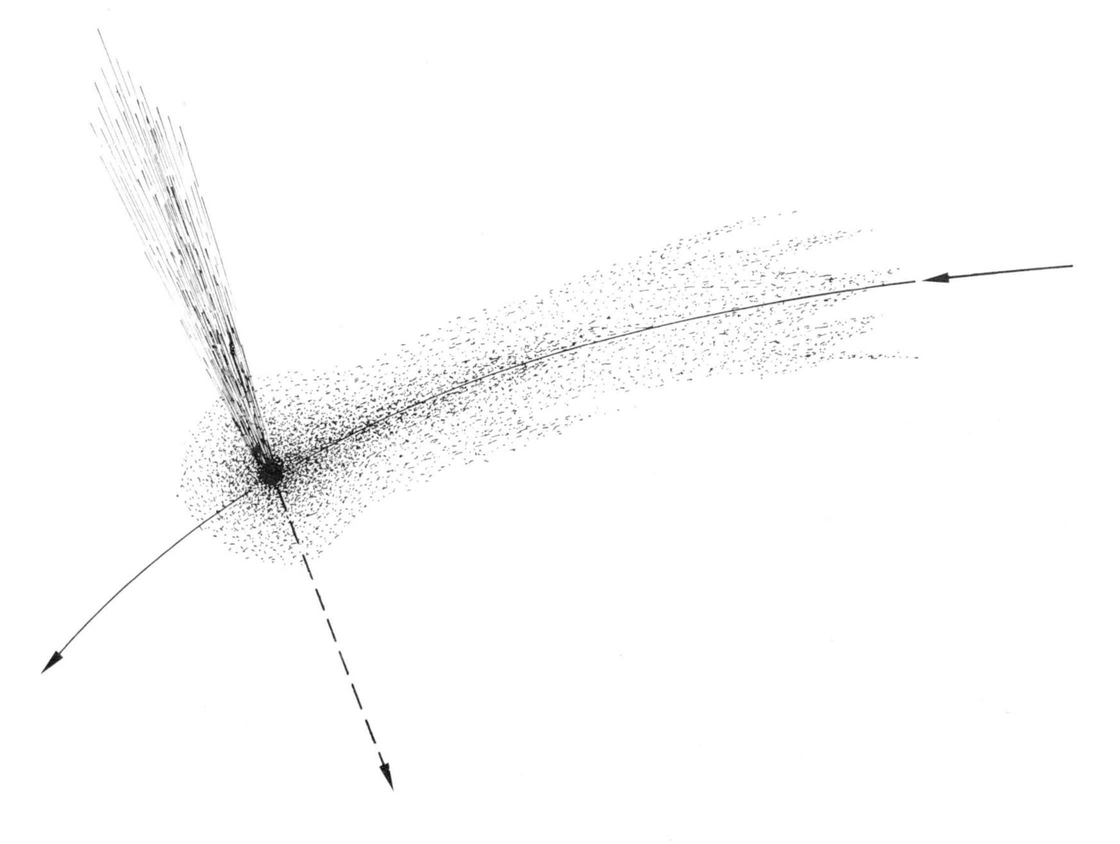

*There are several separate parts to a typical comet. Tails may be made of gas, or dust, or both. The dust is less affected by solar wind pressure, so it tends to form a trail along the orbit, whereas the gas streams away from the Sun. A few comets therefore have two tails, one of gas and the other of dust. Within the comet itself is a small nucleus, possibly made of frozen gas. Surrounding this is the bright coma, which may be due to gases boiling away from the icy nucleus.*

# Meteors: Visitors from Space

Our spaceship Earth cruises round the Sun at a speed of 30 kilometers per second. It collides continually with streams of pebbles, chunks of iron and space dust. Fortunately, we have a superb shield against this barrage. Our life-giving atmosphere also protects us from the missiles of space. A stone, or fragment of iron hitting the upper part of the atmosphere is boiled away to smoke by friction as it whizzes through the air.

Most of the fragments that reach the surface of our planet are no bigger than pinheads. Yet, every day, they add another 1000 tonnes to the mass of the Earth! Several thousand million specks of grit enter the atmosphere daily. A very tiny number of these make trails bright enough to see with the naked eye as they burn up. Perhaps sometime you have seen a shooting star flash through the sky. In fact the shooting stars are high-speed stones.

*This large iron meteorite fell to the ground thousands of years ago near to Grand Canyon, Arizona. It is chained to the ground, but this precaution scarcely seems necessary, since the meteorite weighs almost 400 kilograms.*

*When the Earth passes through a stream of meteoroids, the streaks of light appear to fan out from one part of the sky. The point is called the radiant of the meteor shower. Seen here is the radiant of the Leonid shower, from which several meteor streaks are darting out.*

The bits and pieces of rock drifting through space are called meteoroids. An object that actually crashes right onto the surface of the Earth is a meteorite. The streak of light in the sky that a burning meteoroid makes—the "shooting star"—is just called a meteor. Fireballs are extremely bright meteors. Only a few of these are sighted each year, but if you are lucky enough to see one, it is a splendid spectacle.

The meteoroids in space orbit the Sun just as the planets do. Sometimes the Earth brushes through a great swarm of meteoroids. Most of these swarms travel along the old paths of dead comets. Apparently, the comets leave trails of stones behind them! When the Earth crashes through a swarm, a wonderful display of meteors occurs. In a good shower there may be a meteor about every minute on average. Exceptionally, hundreds may come in an hour. In a meteor shower the bright trails all radiate out from one part of the sky. The table shows the dates when Earth passes through meteor streams. The Leonids gave the greatest display recorded in history in November 1833. In this celestial firework show spectators saw a thousand meteors every minute.

When an average meteorite hits the ground, a few kilograms of fragments will be collected. These are eagerly examined by scientists to see what they are made of. About nine-tenths are stony and one-tenth of them are made of iron and nickel.

Sometimes, very large meteors crash onto the Earth with spectacular results. About 30,000 years ago an iron meteor 25 meters across hit the Arizona desert. It impacted at 15 kilometers per second. An earth shattering explosion left a crater 220 meters deep and $1\frac{1}{4}$ kilometers across. This crater is a major tourist attraction. Fortunately, such large meteorites are extremely rare. Nobody has ever been killed by a meteorite.

There are so many meteoroids in interplanetary space that delicate space instruments must be protected. It is not possible to make a small spacecraft withstand the rare chance of a crash with a large rock, but the fine dust is prevented from causing damage by the use of shielding.

In tropical countries particularly, it is possible to see the dust in interplanetary space. The dust is mainly concentrated in the plane of the solar system, which lies along the zodiac. Sunlight is dimly reflected from this dust. At sunrise or sunset, a cone of faint light can be seen near the horizon and along the zodiac. This is called the "zodiacal light" or "false dawn", and is most readily seen in the tropics on a moonless night.

*A lump of iron as large as an apartment block smashed into the Arizona desert 30,000 years ago and left this crater near the town of Winslow. In the last century, a mining company tried to recover the iron by tunnelling down beneath the crater but they found nothing significant. The Earth has hardly any major meteor craters. Our atmosphere provides a barrier for many meteorites, and the craters that formed millions of years ago have been erased by weather and water. This is not so on the Moon and Mercury, where craters that are billions of years old are still plainly visible.*

# Our Sun

The Sun, our daytime star, is an ordinary member of the starry skies. It is just one single star out of the one hundred thousand million that make up the Milky Way. From a distance of many billions of kilometers, the Sun would look like any other common star. Although there are many strange and unusual stars in the heavens, the Sun is not one of them.

To the plants, animals, and peoples of Earth the Sun is unique because it is near to us. Every living thing on the Earth owes its existence to the fact that the Sun keeps shining and has done so for five thousand million years. The energy we get now from the burning of coal, oil, wood and natural gas was once sun-energy. These fuels are the remains of plants and animals that grew in the warmth of sun-energy millions of years ago. The nearest star apart from the Sun is 300,000 times further away, and the weak star-energy we receive from it cannot possibly replace sun-energy.

Our Sun is far larger than the Earth and also a great deal more massive. A hundred earth-planets placed side by side would stretch from one side of the Sun to the other. Its volume is one million times greater than the Earth and the mass 330,000 times as much.

The distance from Earth to Sun is about 150 million kilometers. Light and heat take eight minutes to race across interplanetary space and reach us from this distance. Although this seems a great separation, only a handful of stars exists within a million times this distance from the solar system. For this reason, we can find out more about the Sun than about any single star.

Exploring the Sun is a major part of astronomy and some observatories look at no other object in the sky. It is essential for us to understand how the Sun makes its heat and how it can affect the weather on the Earth.

The Sun's gravity pulls much harder than the Earth's gravity. A person who could venture to the edge of the Sun would weigh about two tonnes. However, this is an impossible adventure since the Sun has no solid surface and the temperature there is 6000°C. This exceeds the melting temperature of every known substance. The temperature of the surface seems high, but inside the Sun it's much hotter. Its entire globe is a glowing mass of gas. At the center the temperature is sixteen million degrees Centigrade. A grain of sand that hot would fry a person a hundred kilometers away!

As little as a century ago educated people firmly believed that the Sun was nothing more than a flaming ball of fire. A sun made of blazing coal could not, in fact, last even for a million years before becoming a heap of ashen dust. Geologists have shown that the Earth is thousands of millions of years old, and that the Sun has shone throughout that time. In the 1930s physicists showed that the Sun and stars are powered by nuclear reactions.

The gas inside the Sun is three-quarters hydrogen, the lightest gas. Deep inside the hot Sun, hydrogen atoms crowd together. In the jostling a group of them collides so violently with another group that they fuse together and make a completely different substance, helium. In each and every second, 700 million tonnes of hydrogen transform into helium. A small part of this mass of material vanishes in the process and reappears as pure energy. In one second, the Sun's mass falls by four million tonnes. In fifty million years the lost mass is equal to the mass of the Earth.

Flashes of energy burst forth as the hydrogen turns to helium. The great density of matter traps the energy flashes inside the Sun. They wander through the interior for a million years or so before reaching the surface. The energy, then streams off into the blackness of space and within another million years some of it is half way to the nearest great galaxy beyond the Milky Way. If the solar heat processes stopped right now, it would be many thousands of years before the Sun cooled off noticeably.

The Sun is now middle-aged, but can keep the nuclear furnace in action for about another five thousand million years. We need not worry about it's cooling off. In the far future, however, our successors will have to colonize the planets around other stars to escape extinction.

Along with the heat and light, the Sun emits radiation that is harmful to living creatures. Ultraviolet and X-rays damage the cells in plants and animals. Our blanket of atmosphere soaks up almost all of this radiation, although the small amount that reaches the ground on a fine day will make fair skin go darker, or cause painful sunburn if exposure is too long. Astronauts journeying into space have to be protected from the Sun's harmful rays.

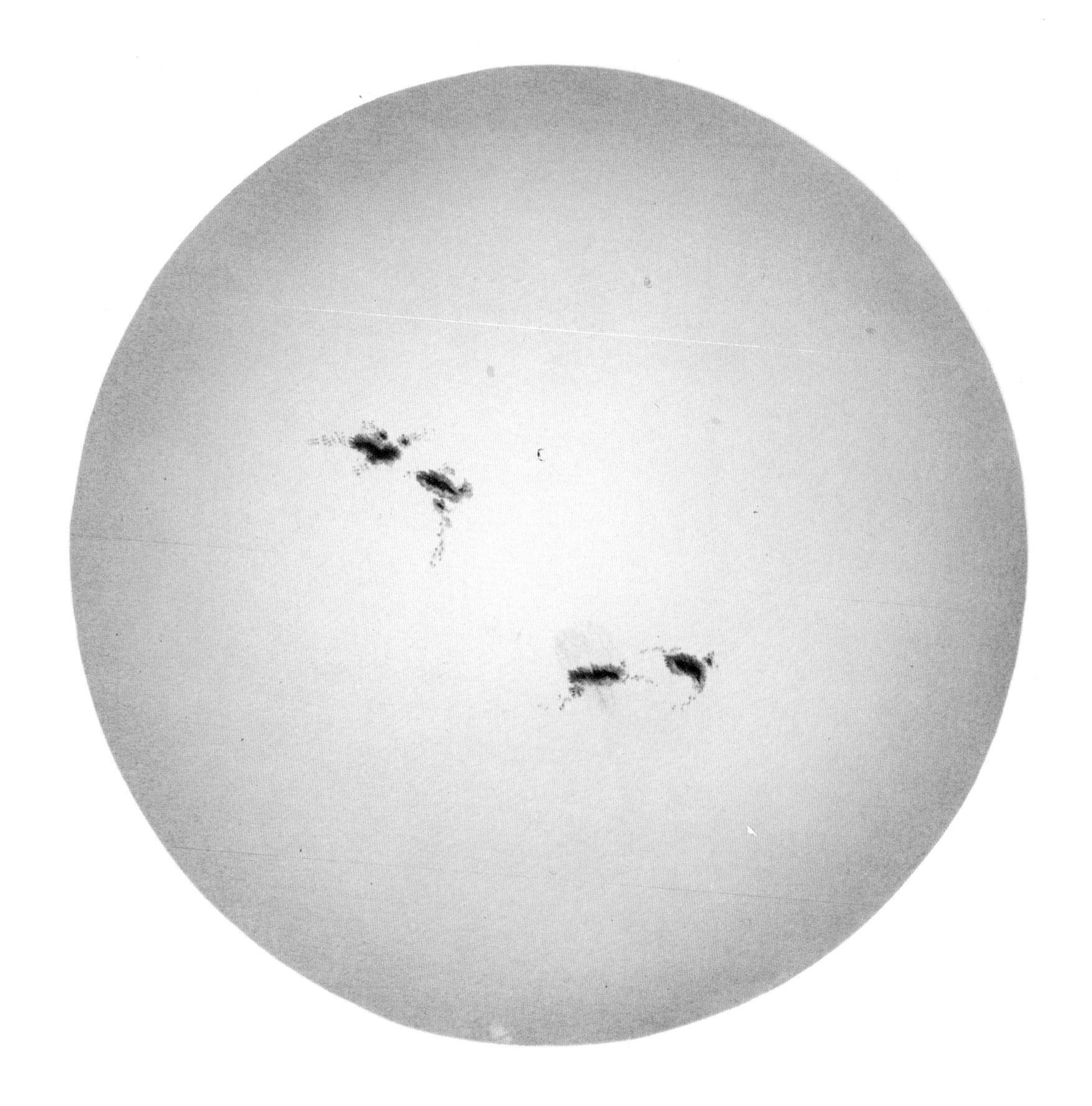

*The Sun is the nearest star and can be observed in far more detail than is possible for any other star. It has a variety of surface features. Dark sunspots come and go; flaming arches of gas leap far into space; and solar flares may flash out brilliantly for a few moments in the active regions. The yellowy surface of the Sun is called the photosphere. It marks the transition between the immensely hot and opaque inside of the Sun, and the thin transparent atmosphere that allows the sun-energy to stream into the blackness of space.*

*This large telescope at the Kitt Peak National Observatory in Arizona is used by solar astronomers. The diagonal tower runs underground for a considerable distance, so the telescope is actually larger than the structure seen here.*

The Sun's blinding light will hurt your eyes permanently if you stare at it. **No one should ever look at the Sun through any type of magnifier, field glasses or a telescope.** The cheap filters sometimes sold with small telescopes are likely to crack and they may not cut out the dangerous invisible rays. Take our advice: destroy any solar filter supplied with the telescope so that no one will be tempted to use it.

Even astronomers do not look through their telescopes directly at the Sun. Special instruments fitted to solar telescopes allow them to follow the behavior of the Sun. Information gathered by solar observatories may one day aid long-term prediction of the weather.

# Spots on the Sun

The Sun is far too hot to have any kind of solid surface, and the yellowy disk that we see is actually the topmost layer of glowing gas. Above this layer is the cool zone of transparent gas called the chromosphere. Solar astronomers use photographs taken with special telescopes to study the gaseous surface. Frequently, the photographs are made with filters that isolate the light associated with one particular type of atom, such as hydrogen or calcium. In this way it is possible to select for research individual parts of the Sun's surface.

A photograph taken under very good observing conditions shows that the Sun's surface has a mottled or bubbly appearance. Glowing speckles of gas jostle for attention, with darker spaces snaking between them. New flecks constantly appear, and then, after a few moments, dissolve away again. Each of these bubbles is generally as large as the British Isles. They are, in fact, gushers of hot gas shooting up from just inside the Sun. The yellowy layer is somewhat like the surface of a gently boiling pan of milk.

Dark blotches called sunspots were noticed even in ancient times. The largest ones may stretch across one-third of the visible disk, which is an enormous thirty times greater than Earth's diameter. These giant sunspots are rare, but they can be seen by the naked eye *at sunset* if they are present at all.

Do not try to look at the Sun or its sunspots with any magnifier or telescope. If you have a simple telescope with a sturdy tripod then you can try to project an image of the Sun on to a piece of white card. With a little practice you can learn how to make a sharp image of the Sun in this way. Sunspots will appear as greyish specks. If a record is kept of the positions of the spots on several successive days you will soon see that they change in size and shape, and that the Sun itself slowly rotates.

Sunspots look like holes in the fiery surface of the Sun. In fact they are areas which are about 2000°C cooler than the surrounding surface. This makes their temperature roughly 4000°C. Now something that hot is actually extremely brilliant: sunspots only *look* dark because they are cooler and dimmer than the rest of the Sun. If one could be plucked from the Sun and examined separately it would seem as bright as the full Moon.

How are sunspots caused? The Sun's very strong magnetism probably bursts out from the inside of the Sun, and a pair of spots is created at those two points where the tangle of magnetism leaves and subsequently re-enters the Sun. Small spots vanish in a matter of hours but the larger ones may remain for several weeks before finally disintegrating. Small spots are 3000 kilometers across, most spots are more or less as big as the Earth, and huge spots span 150,000 kilometers.

By watching the spots, we can see that the Sun rotates. As seen from Earth, the spots take about 26 days to make one complete circuit near to the equator, whereas near the

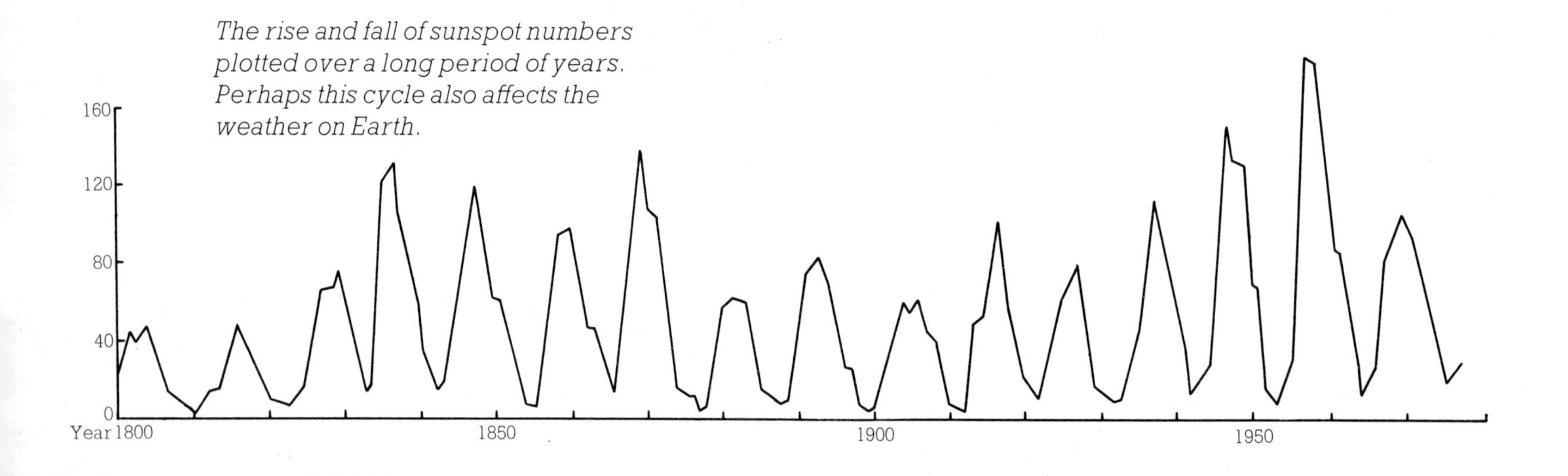


*The rise and fall of sunspot numbers plotted over a long period of years. Perhaps this cycle also affects the weather on Earth.*

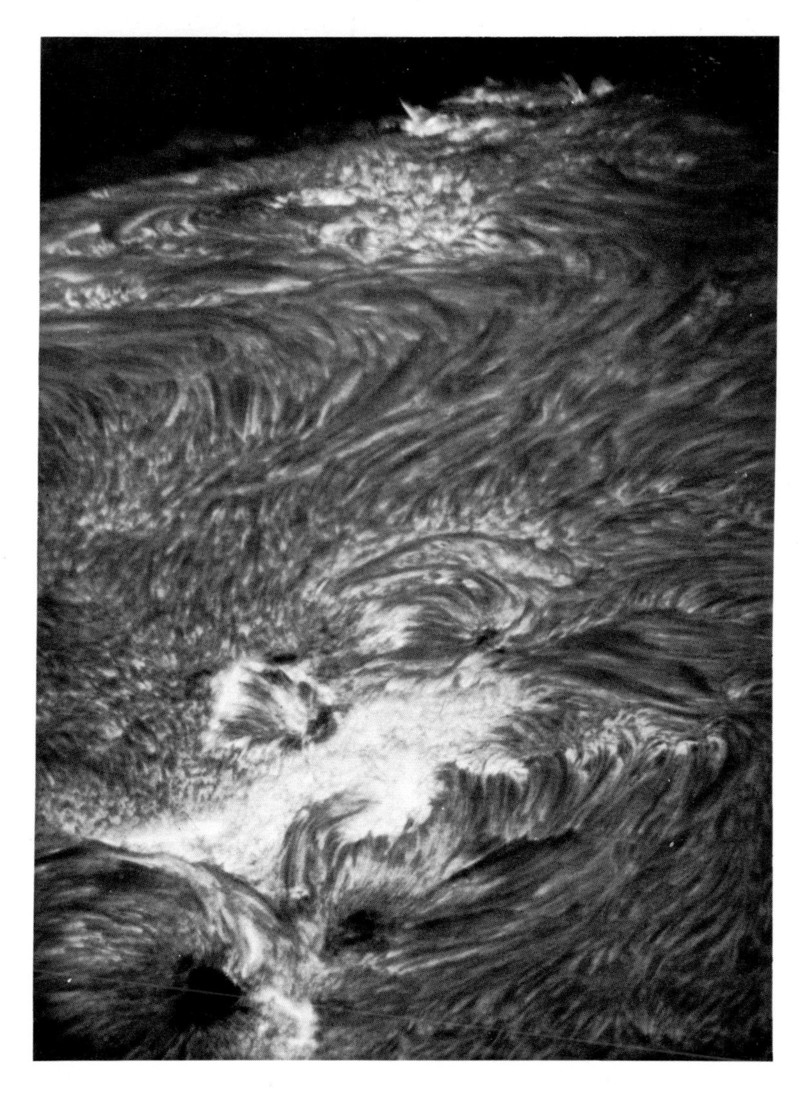

*A solar storm rages close to sunspots, producing flares of white light.*

poles they need 40 days. This difference in rotation proves that the Sun cannot be a solid body like Earth: if it were, all the spots would be carried round in the same time. If you try keeping track of spots it's fun to see if one is still there two or three weeks after vanishing round the back.

Records of sunspots for the last three hundred years show that the number of spots and their size varies on a cycle lasting for around eleven years. The number of spots steadily increases for five or six years; then there is a decline in numbers for the following four of five years. At the end of the cycle, there may be no spots at all for many months until the new cycle gets going. Vast changes in the Sun's magnetism probably cause this variation in the sunspot numbers. Many other aspects of the Sun's behavior change through the cycle, and this probably affects in turn the weather on Earth.

Sunspots can interfere with radio communications. Electric particles are flung into space by flares and eruptions near the sunspots. The solar electricity stirred up in these storms changes the upper part of our atmosphere. At such times, fadeout of long-range radio signals may be noticed.

The chromosphere, the cool layer of atmosphere above the yellow surface, can only be seen readily during total eclipses. The temperature in this thin layer is 4500 degrees. Above the chromosphere we come to the intensely hot and invisible corona, where the temperature soars to an amazing one million degrees Centigrade. The gas in the corona is boiling away into space. The rush of gas is termed the solar wind.

*Simon Mitton projects the image of the Sun onto a piece of card held beyond the eyepiece of a small telescope. Even on the lowest magnification small spots can be seen clearly if you use this safe method to view the Sun. A shield of cardboard resting on the eyepiece is used to cast a shadow, and thus cut down the stray light. The finder telescope has been removed from the main tube of the telescope in order to avoid the temptation of looking through it while lining up the telescope.*

# Action on the Sun

Sometimes great storms erupt in the Sun's atmosphere. Extra-hot gases swirl and boil up from below and then stream high above the surface. Near to sunspots, a type of solar storm known as a flare may burst out. A flare is a spectacular release of energy in the chromosphere. A vast region switches on and off in a series of spectacular flashes, each one like an enormous lightning stroke. In only five minutes, the flare spends all its energy and then everything is over as far as the Sun is concerned.

The flaring activity flings large clouds of electric particles away from the Sun. These travel at millions of kilometers per hour. After about two days, the particles reach the vicinity of the Earth. Some of them are funnelled in towards the polar regions of the Earth. This is because the Earth is magnetised in the same manner as a giant bar magnet and the electric particles from the Sun are guided by the magnetism. The magnetism is strongest at the north and south poles. As the electric particles are drawn down into these regions, they strike the gases in the upper part of the atmosphere. This makes the air glow and emit the beautiful light displays known as the aurora.

The finest auroral displays take place two years after maximum sunspot activity. When the aurora is very active, it may be possible to see it from southern Europe or the southern United States. Normally, however, it is only seen further north. A flight in a passenger aircraft across the Atlantic Ocean at night may provide a chance to see stunning displays as the aircraft passes close to Greenland. In Australia the southern aurora is rarely seen.

Earth is able to shield us from the worst blasts of high-speed particles ejected from the angry face of the Sun. Our tiny planet protects us with an invisible magnetic shield. Magnetism forms a cage around the Earth that deflects most of the particles or funnels them in at the poles. Inside the magnetic cage two doughnut-shaped compartments can trap electric particles. These rings are named the Van Allen Belts after their discoverer, James Van Allen

During a total eclipse another type of solar disturbance is seen. Fantastic plumes of gas arch their way above the Sun, reaching altitudes of hundreds of thousands of kilometers before crashing back to the surface. These are the prominences and they can be seen by special solar telescopes at any time.

X-ray telescopes on satellites can photograph the corona of the Sun. At its temperature of one million degrees Centigrade the corona emits X-rays rather than visible light. The corona has tangles of magnetic field running through it. These zones of extra magnetism may show as bright spots on X-ray photographs.

*An enormous solar flare, photographed during the American Skylab missions, spews out into the blackness of space.*

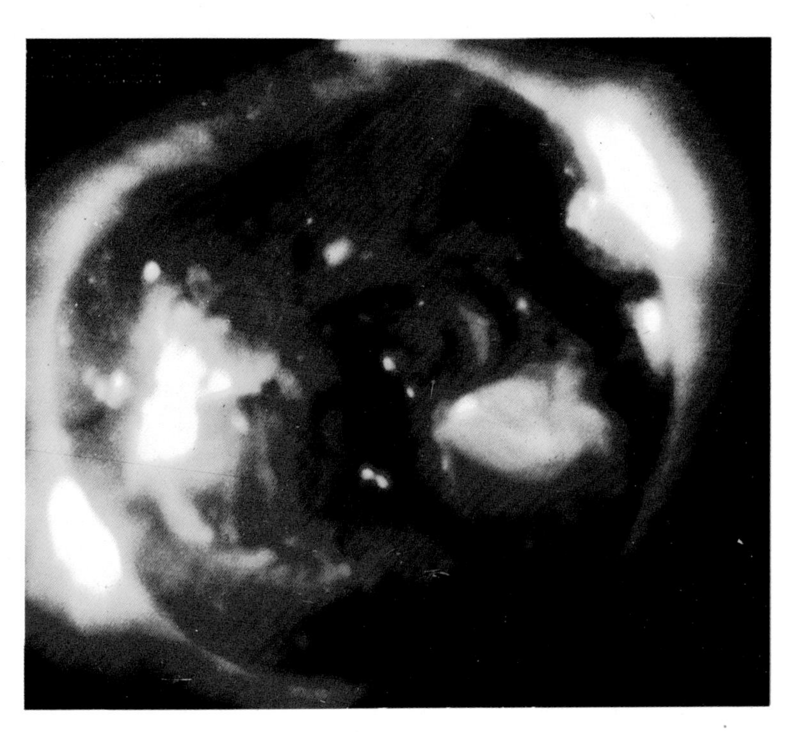

*This X-ray image shows the uneven structure of the Sun's outermost layers.*

*Beautiful auroral displays are a direct result of the collision of particles from the Sun with the upper layers of our atmosphere.*

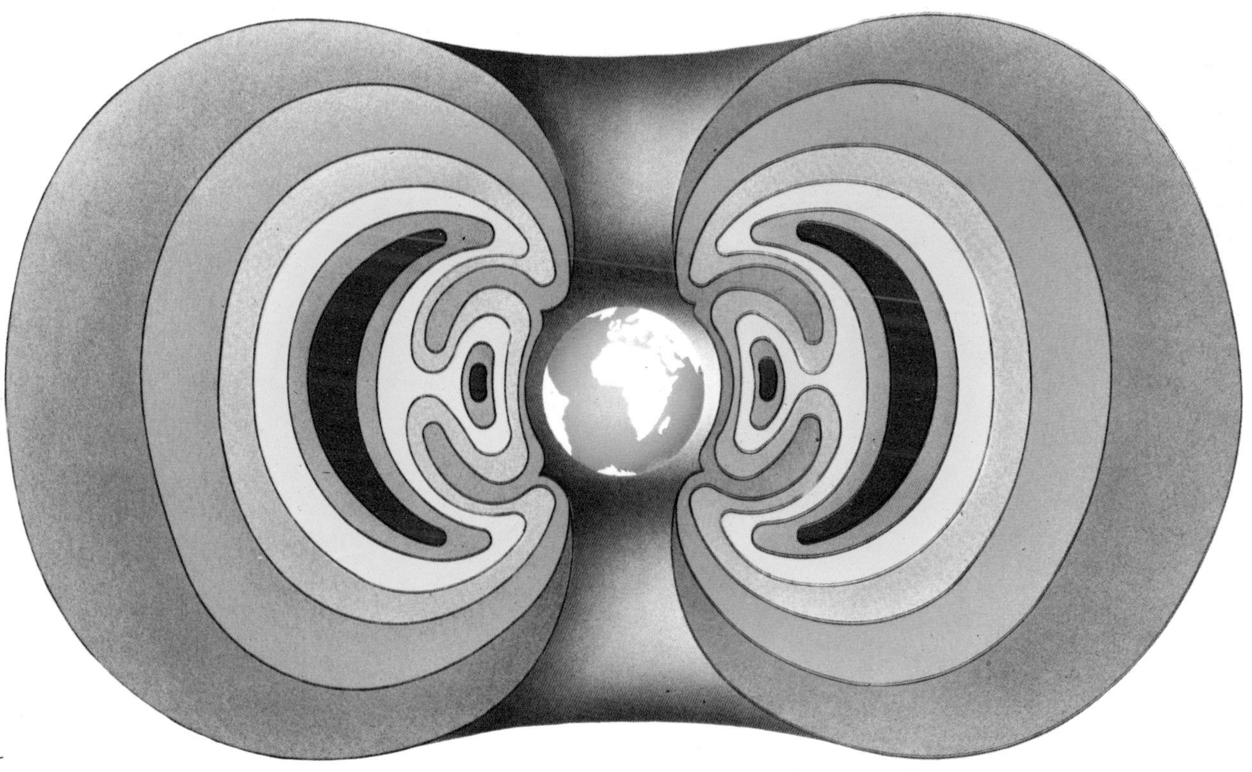

*Earth's magnetic cage, or magnetosphere, protects us from deadly blasts of solar particles.*

# The Stars

Each twinkling point of light in the night sky is really a distant sun. Our Sun seems very special to us because we depend on it for light and warmth. In fact, it is only one very ordinary star in a vast population. Even through a telescope the stars still appear merely as points of light, although with a telescope many more stars can be detected than can be seen by eye alone. But the stars are so much further away from us than the Sun that even the world's largest telescopes are unable to discern the surface features of the nearest ordinary stars. Only in the case of a handful of giant stars can the telescopes vaguely make out the disk of the star. Sometimes star images in telescopes, or on photographs, do indeed look like disks, but these are distortions to the true point image caused by the telescope and the Earth's atmosphere.

Since it's so hard to "see" any star, all of them must be at great distances compared to the Sun and planets. The nearest star to our solar system. Proxima Centauri, is over four light years away. A light year, of course, is the distance covered by a ray of light in one year. Since light travels at 300,000 kilometers every second there are ten million million kilometers in a light year. The light year is a useful unit for marking out large distances in astronomy.

The distances to the nearer stars are found by watching the way in which they change their positions slightly against the background of more distant stars during the course of the year. This relative change of position is called parallax. You can see the effect for yourself as you speed along in a bus or train. Objects close to you flash past the window, whereas objects further away can be seen for some time. The parallax effect causes the nearby stars to nod back and forth across the sky as the Earth makes its annual journey round the Sun. The parallax method only works for finding distances out to one hundred light years. Beyond that distance, astronomers have devised ingenious methods for finding distance. Many of these involve a careful analysis of the light from the star to determine its properties. In particular, if the true output of light from a star can be deduced then a measurement of the quantity of light that we actually receive from the star tells us how far the light has travelled and spread out.

The stars are not spread out evenly throughout all of space. Our Sun and all the stars that can be seen by eye belong to a giant disk-shaped collection of a hundred thousand million stars. This giant family of stars is our Galaxy, 100,000 light years (one million million million kilometers) from side to side. Beyond the edge of our Galaxy, space is totally devoid of stars until the nearest other galaxies are reached.

Some of the stars in the Galaxy are alone, like our Sun, but many are linked in double, triple, and quadruple systems. Stars also form larger groupings called clusters.

*The globular cluster in the constellation Centaurus. This swarm of stars is thousands of millions of years old.*

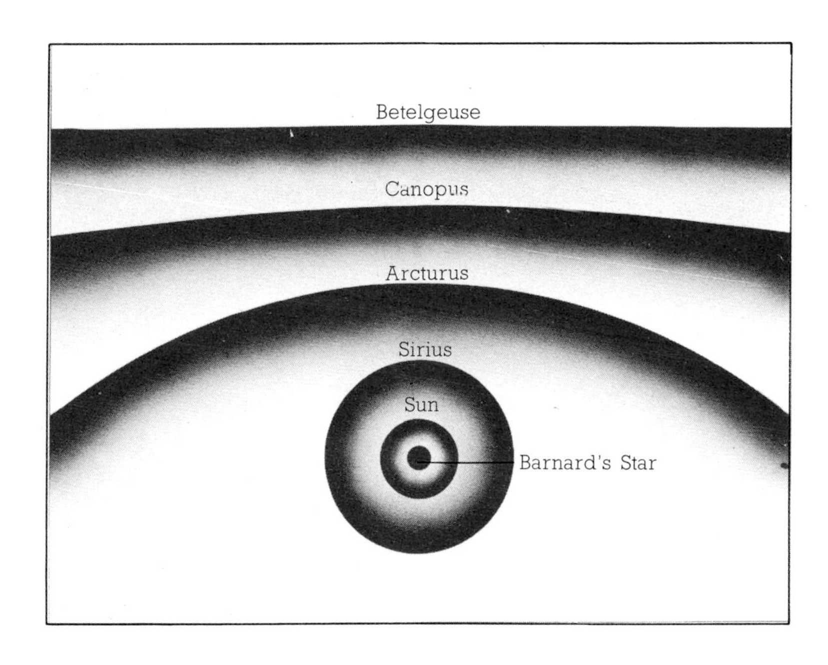


*The relative sizes of some well-known stars are compared. Supergiant stars are so large that they could swallow up the inner solar system.*

Two quite distinct kinds of cluster are found: open clusters and globular clusters. The open clusters are rather loose collections of stars, about one thousand in all. Many individual stars can be picked out in an open cluster. For example the Pleiades (Seven Sisters) in the constellation Taurus is a good group for you to find. Several of the stars can be seen without a telescope, although the misty patch of light becomes a hundred twinkling stars with field glasses. Globular clusters are different from open clusters. They are tightly-packed balls of stars, with up to a million members. In this swarm, only stars at the edge can be picked out individually. Two globular clusters, one in Centaurus and the other in Tucana, may be seen as hazy blobs in the southern skies.

Stars do not appear to be equally bright in the night sky. The range of star brightnesses is partly due to the fact that the stars are scattered and not all at the same distance from us, and partly because all stars are not giving out the same amount of light. Stars cover a whole range of sizes, temperatures and masses, and they vary in their colors too. The most massive stars are the hottest and they are large, shining blue-white. The yellow Sun is a smallish star, called a dwarf. Cooler, redder stars are even smaller than is the Sun. The giants among the stars have puffed out to become many times larger than they once were. Really immense stars are called supergiants.

The patterns of the stars in the sky will not change noticeably in a lifetime, yet the stars are in motion. Over many thousands of years the constellation shapes do change. Yet we can barely detect the star movements even with telescopes because they are so far away. Very careful measurements have to be made on photographs taken a few years apart. Such measurements show that our starry Galaxy is a great swirling mass, a wheel of stars turning slowly in space.

*The open star clusters in the constellation Perseus.*

# The Message of Starlight

How insignificant each speck of starlight seems against the blackness of the night! Yet, each tiny beam is packed with information about its sender. Astronomers have learned how to read the messages in starlight. Starlight tells us what the stars are made of, how big and hot they are, and how they are moving.

Before the information in starlight can be understood, it has to be decoded. First of all the light from a star is collected and concentrated by a telescope. Next, a special instrument on the telescope analyzes the light. The instrument is called a spectrograph. Finally, the result is stored as a photograph or on computer tape.

*The telescope that took this photograph of a cluster of stars had a prism placed across the front. Instead of appearing as a point of light, each star is drawn out into spectrum. Look closely and you can see fine dark absorption lines crossing each tiny spectrum.*

*The spectra of the Sun and stars are crossed bv many narrow, dark absorption lines. They are often called Fraunhofer lines, after the man who first listed some of them in the Sun's spectrum. The study of these lines helps astronomers understand how stars work and what they are made of.*

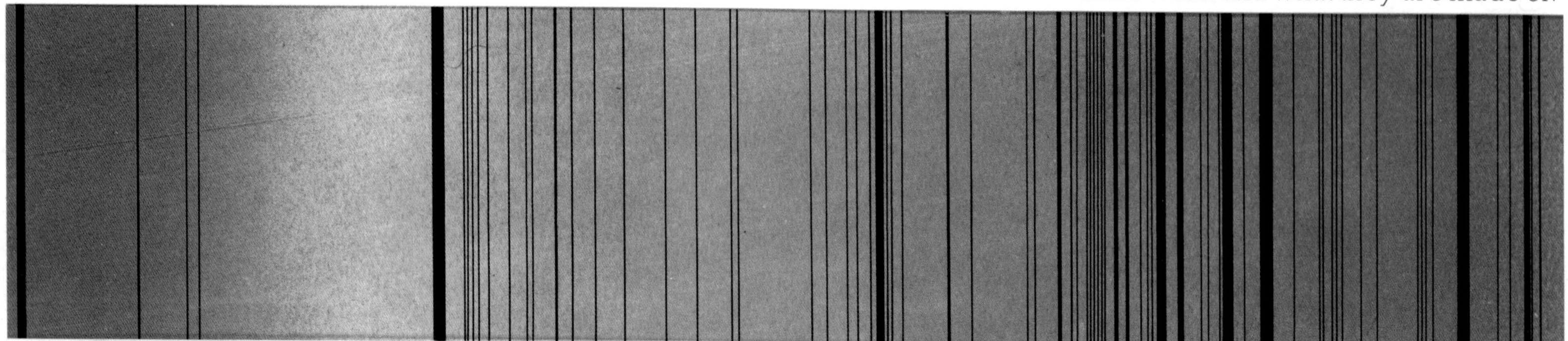

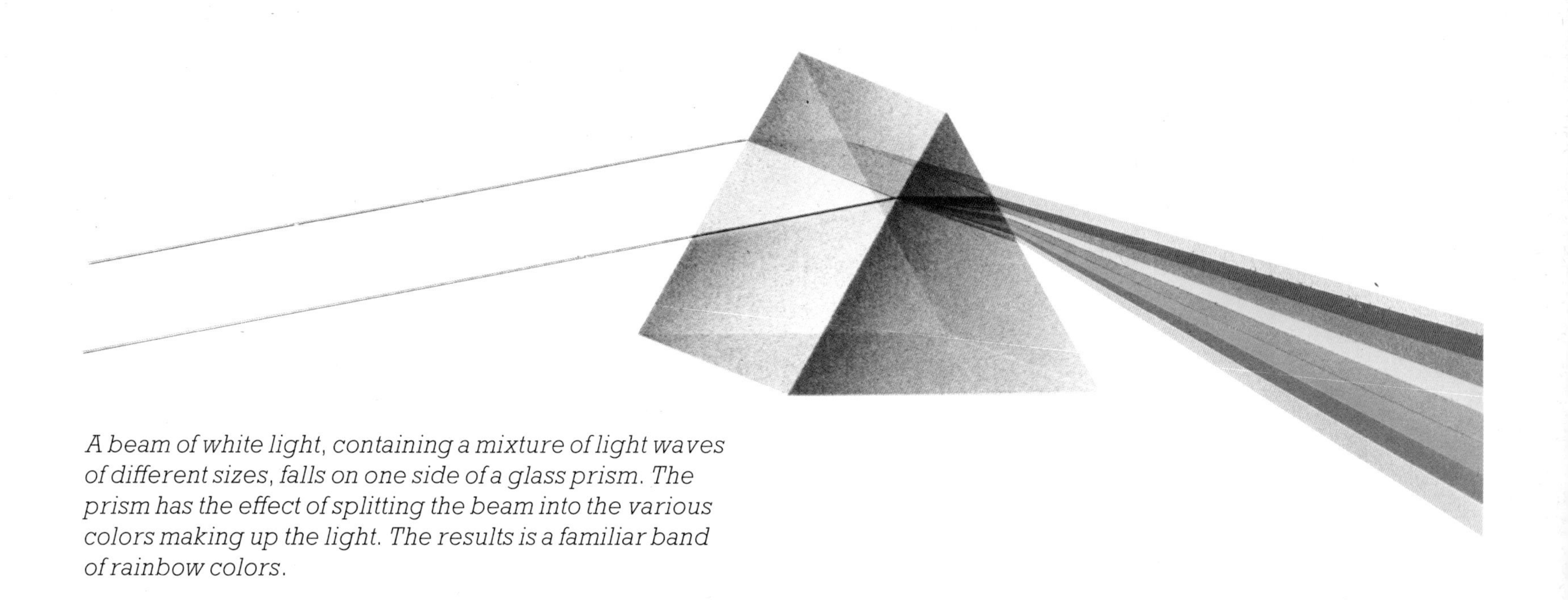

*A beam of white light, containing a mixture of light waves of different sizes, falls on one side of a glass prism. The prism has the effect of splitting the beam into the various colors making up the light. The results is a familiar band of rainbow colors.*

Light is often described as consisting of "light waves". Light does indeed travel in the form of waves, but these waves are so small that thousands fit into a millimeter. Light waves are energy waves and they can travel through the total emptiness of space. Animals use their eyes to detect light waves. We can see things that give out their own light energy, such as the Sun, the stars or an electric light bulb. We can also see objects that reflect light, such as our own bodies and the planets. Light waves come in a whole range of sizes. Some are longer, some shorter. We can tell the difference between light waves of different sizes because we see them as different colors.

Different mixtures of light waves give us an infinity of shades from which to choose. The hues of the rainbow show how waves of different sizes have different colors. Red light has the longest waves, and purple light the shortest. A rainbow is caused when the Sun's light shines through falling raindrops. The blobs of water are able to split the sunlight up into its range of colors—into what is called a spectrum.

A spectrum can be made from any beam of light, without needing rain. A glass prism in the path of the light beam will split it up into its rainbow colors. Another way of getting a spectrum is to use a piece of flat glass with many lines engraved close together, which is called a grating.

Inside an astronomer's spectrograph, a prism or a grating splits the starlight into its range of colors. The colors are in varying proportions from the different kinds of stars. Cool stars, for example, are sending out more red light than blue or purple. That is why they look reddish in the sky. Hot stars emit more evenly through the spectrum. All the colors combined look white or bluish.

At some particular colors there are narrower gaps in a star's spectrum where there is very little light. The gaps look like black lines cutting across the continuous background rainbow. They are called absorption lines. These lines are very important because the number and positions of the lines reveal what gases are present in the stars. Different gases absorb different colors as the light leaves a star. Each type of gas puts its own "fingerprint" of black lines into the spectrum, and this allows particular gases to be identified. From the study of absorption lines we have learned that stars are made mainly of hydrogen. The rest of the material, about a quarter, is helium. Many other elements such as oxygen, silicon, iron, and nickel make up only one per cent of a normal star.

Human beings have evolved on the Earth, bathed in sunlight. As a result, our eyes are most sensitive to the range of light colors emitted by the Sun. However, there are other kinds of "light" which our eyes cannot detect. Each kind of radiation has been given a different name, but they are all really like light, except that each has waves of a different size. Gamma-rays, X-rays, ultraviolet, infrared, microwave and radio waves, are all part of the family of electromagnetic radiation. All these types of rays can be picked up from various objects in space, but special telescopes have to be used. Our eyes are no good outside the range of visible light.

# Double Stars and Variable Stars

Our Sun is a solitary star, over four light years from its nearest neighbor, but many of the stars in the sky are actually double. The two stars that make up a double star, or binary star, are held together by the pull of gravity between them. The planets in the solar system travel in orbits around the Sun each held by the Sun's gravity. In a binary star, each member is in orbit around the balancing point between the two. Closely-spaced pairs of stars may take only a day or two to complete their circuits. Distant pairs may take over a hundred years!

Some double stars can be detected easily with a telescope. Both stars can be seen, and over a period of time, one seems to travel around the other. But only the pairs separated by large distances are directly observable. Close pairs cannot be split apart into two star images, even in the largest telescopes. However, astronomers can tell when there are actually two stars very close together by looking at a spectrum. As the stars circuit their orbits, there is a regular cycle of changes in the spectrum. From the study of this spectrum it is possible to find out what each individual star is like.

As well as doubles, there are star systems with three or even more members, though not many are known. A famous example is Castor in the constellation Gemini. There are six stars altogether in this multiple system. Three stars can be detected with a telescope, but each, in turn, is a close double.

The orbits of some double stars are lined up so that from Earth each star disappears behind the other one as they move around in their orbits. Normally, we can see the light from both stars together, but when one of them is hidden, there is a sudden dip in the amount of light we see. The starlight gets fainter for a short while, then returns to its original brightness. Pairs of stars like this are called eclipsing binaries.

There is a well-known eclipsing binary in the constellation Perseus. It is called Algol. Every 69 hours the brightness of the Algol double star drops sharply by more than a whole magnitude for the space of a few hours.

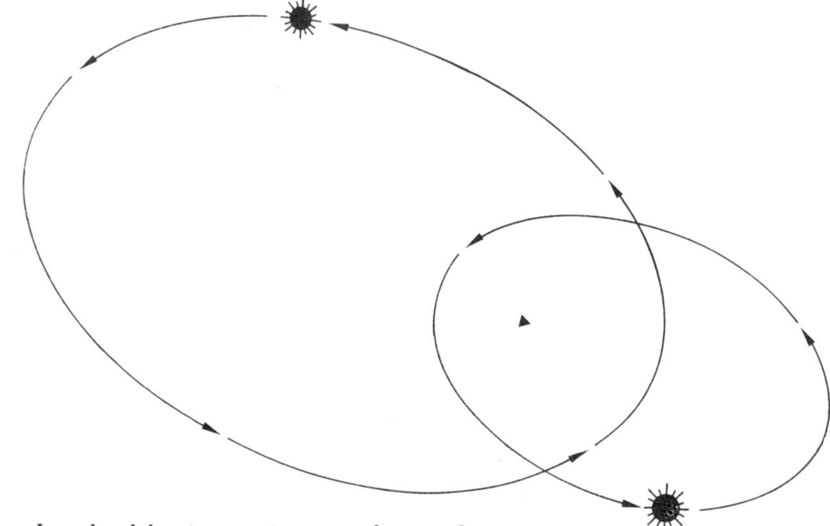

*In a double star system, each member travels in its own orbit around the balancing point between them. Each orbit is an ellipse. The two stars always keep on opposite sides of the balance point as they swing around, just like a pair of ice skaters.*

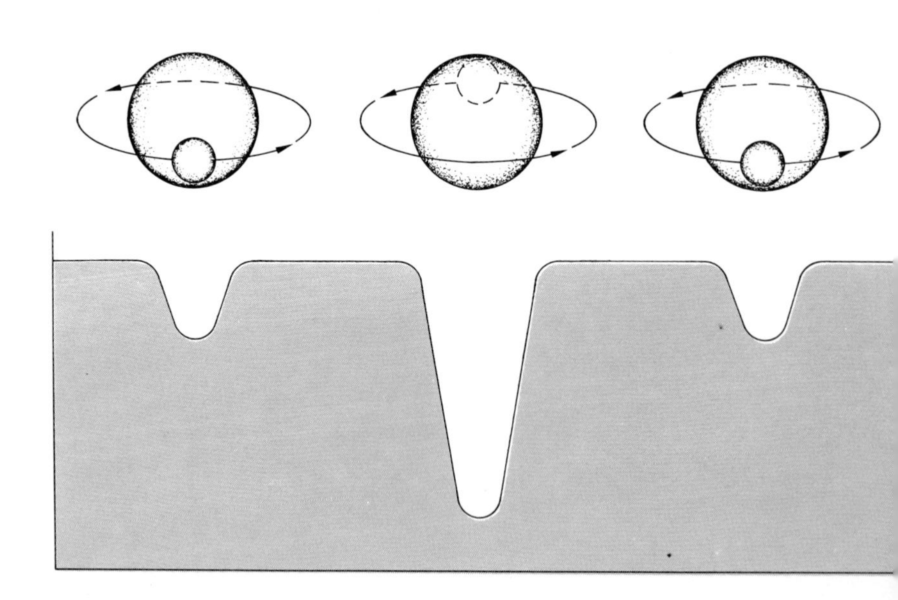

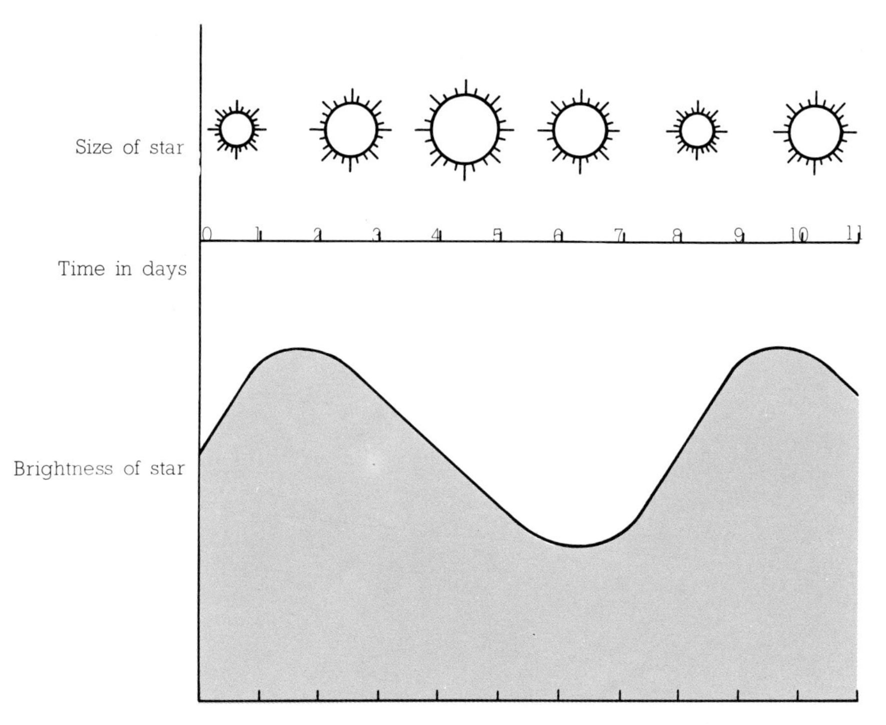


*This variable star puffs in and out with a regular cycle of a few days. As its size changes, its brightness changes too.*

Eclipsing stars are not the only ones whose light output varies. There are many other kinds of variable stars. Some of them change in a very regular way. The Cepheid variables, named after a star in the constellation of Cepheus, are an example. These stars actually swell and shrink quite regularly. As they "breathe" in and out, their brightness rises and falls. Other variables are less well-behaved. Some that are usually dim unexpectedly flare up every now and again, gradually fading to their former level after each outburst. Others do the opposite and without any warning decide to become fainter than they usually are. The variables are older, giant stars. They are well on into the later part of their life-time. The forces that normally keep stars like the Sun shining steadily have got out of balance, causing the stars to be uncertain about their brightness.

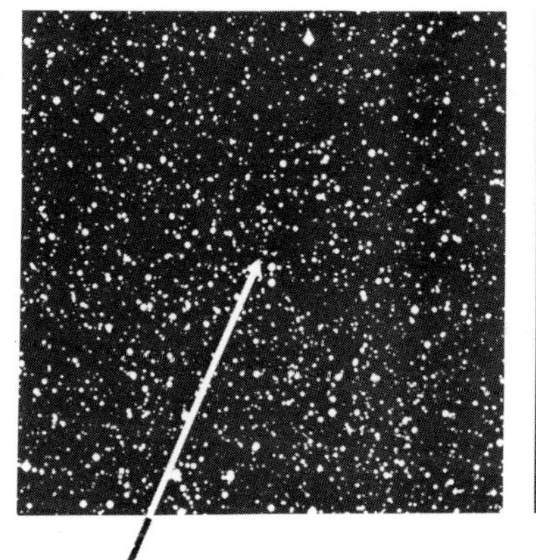

*A nova is a star that suddenly flares up to be many magnitudes brighter than it was. A really brilliant nova may have been a star which was hardly detectable in a large telescope. Then, suddenly, it becomes visible to the naked eye. Here are two photographs taken before and after the nova's outburst. A nova gradually fades again after the spectacular event. Nova means "new star".*

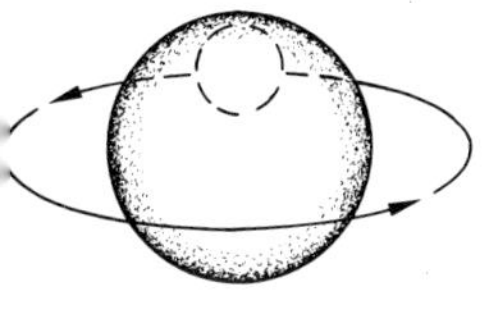

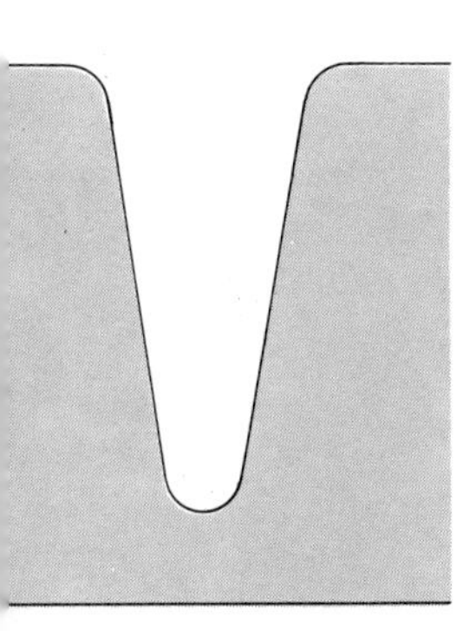

*The members of an eclipsing binary system pass each behind the other as they move in their orbits. The amount of light that we can see dips during each eclipse. In this example the two stars have different sizes and brightnesses. The biggest dip in brightness comes when the smaller, fainter star blots out part of the larger and more luminous star.*

# The Life of a Star

Over the human lifespan, the countless stars of the Milky Way seem practically changeless. Occasionally, a nova flares suddenly to change the pattern of a familiar constellation for a few weeks, until once again it has faded into insignificance. Rarely, even a supernova blazes forth like the one in AD 1054 that left behind the Crab Nebula. Variable stars flicker with uncertain light.

The stars do eventually alter. Nothing in the Universe lasts forever. A coal fire dies when the last embers turn to ash. A star dies when its huge store of nuclear fuel is finally exhausted. Even today old stars are fading out, while new ones are being born to replace them. We can see stars in all stages of evolution, from infancy to old age.

Very young stars are found still embedded in the gas from which they form. The first faltering light from brand new stars has been seen in the Orion nebula, for example. Our own Sun is somewhere comfortably settled in middle age. Some of the oldest stars that are known are in the globular clusters.

Perhaps you wonder how it is possible to work out the ages of the stars. Nobody can follow the progress of a single star from birth to death. However, imagine that you have never before seen a tree and are suddenly transported to the middle of a forest. There you will see trees in every state of development from seedlings to gnarled giants. Knowing a little biology, you would soon work out the life cycle of a tree. In a similar way, from the laws of physics and observations of stars of different kinds, astronomers have been able to deduce the sequence of events in the life of a star.

After a star has formed, it soon settles down to a steady existence. Nuclear reactions in its innermost core conver hydrogen to helium, releasing energy at the same time. Eventually, all the hydrogen in the interior is consumed. Changes then take place in the star's internal balance. The outer layers puff out to give the star giant proportions, while new reactions, working on the helium, start up inside. More changes occur and the star will go through a phase of being variable. Ultimately, there is no possible source of energy left. Smaller stars shrink into white dwarfs. Massive stars blow up as supernovae. The material blasted out by a supernova becomes part of the interstellar gas, the birthplace of a new generation of stars.

One of the last stages in the lives of stars, before they become white dwarfs, produces some of the most fascinating objects in the sky. These are the planetary nebulae. Their regular shapes and beautiful colors make them very attractive. (In reality they are nothing to do with planets. The name is a relic from early telescopic observers who thought their disk shapes looked similar to planets.) A planetary nebula is formed when the star at the centre throws off a layer of itself. The shell of gas travels outwards, like a smoke ring, but in three dimensions.

*Changes in the course of a star's lifetime, from birth in a cloud of gas, to old age as a white dwarf.*

*Three planetary nebulae, showing the different forms they may take. These are the Ring, Dumbbell and NGC 6302.*

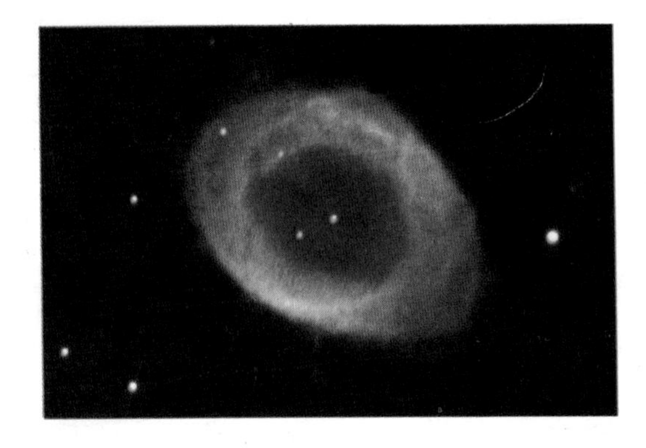

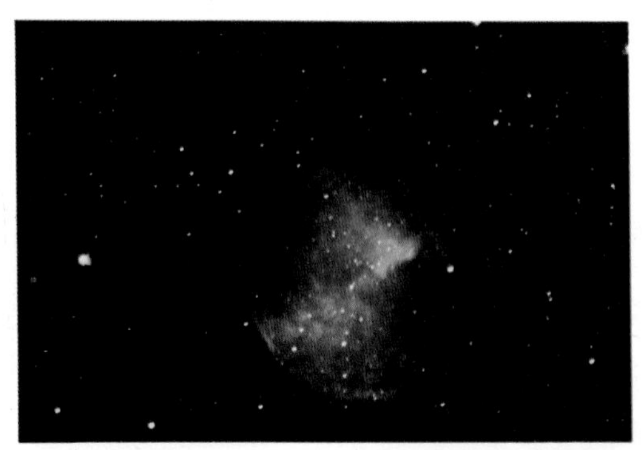

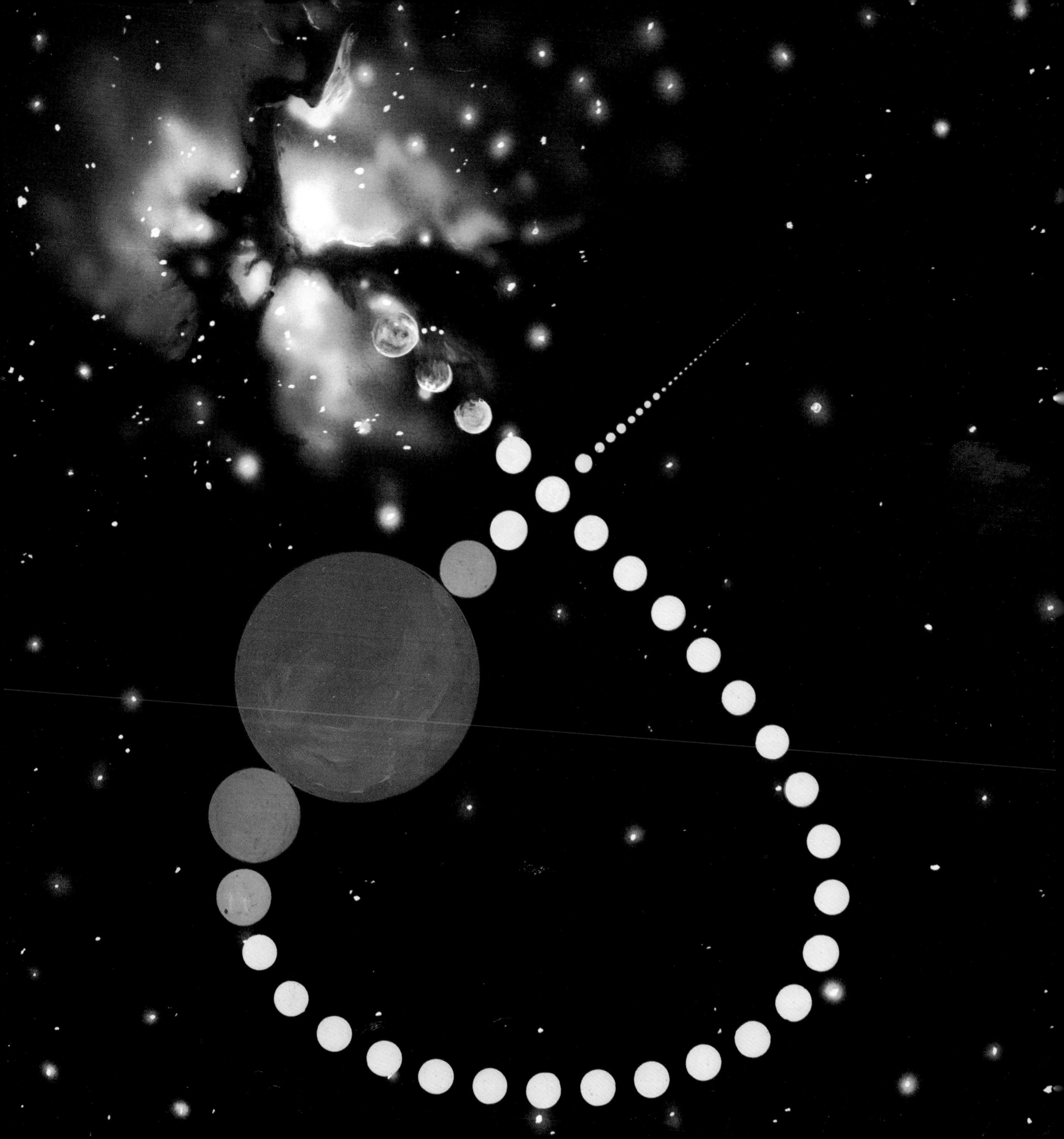

# Star Deaths

Throughout its life a star is a battle arena. The force of gravity tries to crush the star in on itself, but this inward shrinking is resisted by the outward pressure of the star's material. Eventually, however, the star becomes exhausted. Gravity takes control and the star takes on a form quite unlike that of normal healthy stars such as the Sun. A large star may even disappear altogether by turning into a black hole lost in space.

The force of gravity always attracts: it wants to pull particles of matter closer together all the time. We humans experience weight because the mass of the Earth tugs at the mass of our body and we feel a force. Every atom of the body is attracting all other atoms by gravity. Since a normal star is very massive, perhaps being a million times more massive than Earth, its internal gravitation is high. Try to imagine what it's like inside the Sun: at one-tenth of the distance from the outside to the core the pressure is already a million times higher than atmospheric pressure at the surface of the Earth. Halfway in it rises to a thousand million atmospheres. This crush is resisted by the pressure exerted by the hot gas inside the Sun. The gas is kept heated by the nuclear furnace.

When the nuclear fires finally dwindle, the star gas cools and then gravity is in command. What happens at this stage depends on the mass of the star.

A dying star similar to the Sun collapses until it is about as large as the Earth. No really spectacular explosion occurs. It just falls into a heap of radioactive ash and gently flickers out. It has turned into a white dwarf star. A cupful of white dwarf matter weighs one hundred tonnes!

If a star is somewhat more massive than the Sun, the inward plunge pushes it past the white dwarf stage. The collapse does not stop until the star is ten kilometers across. At this point it is a dense ball of nuclear particles. A cupful of neutron star material weighs nearly a million million tonnes! Some neutron stars spin rapidly and emit a

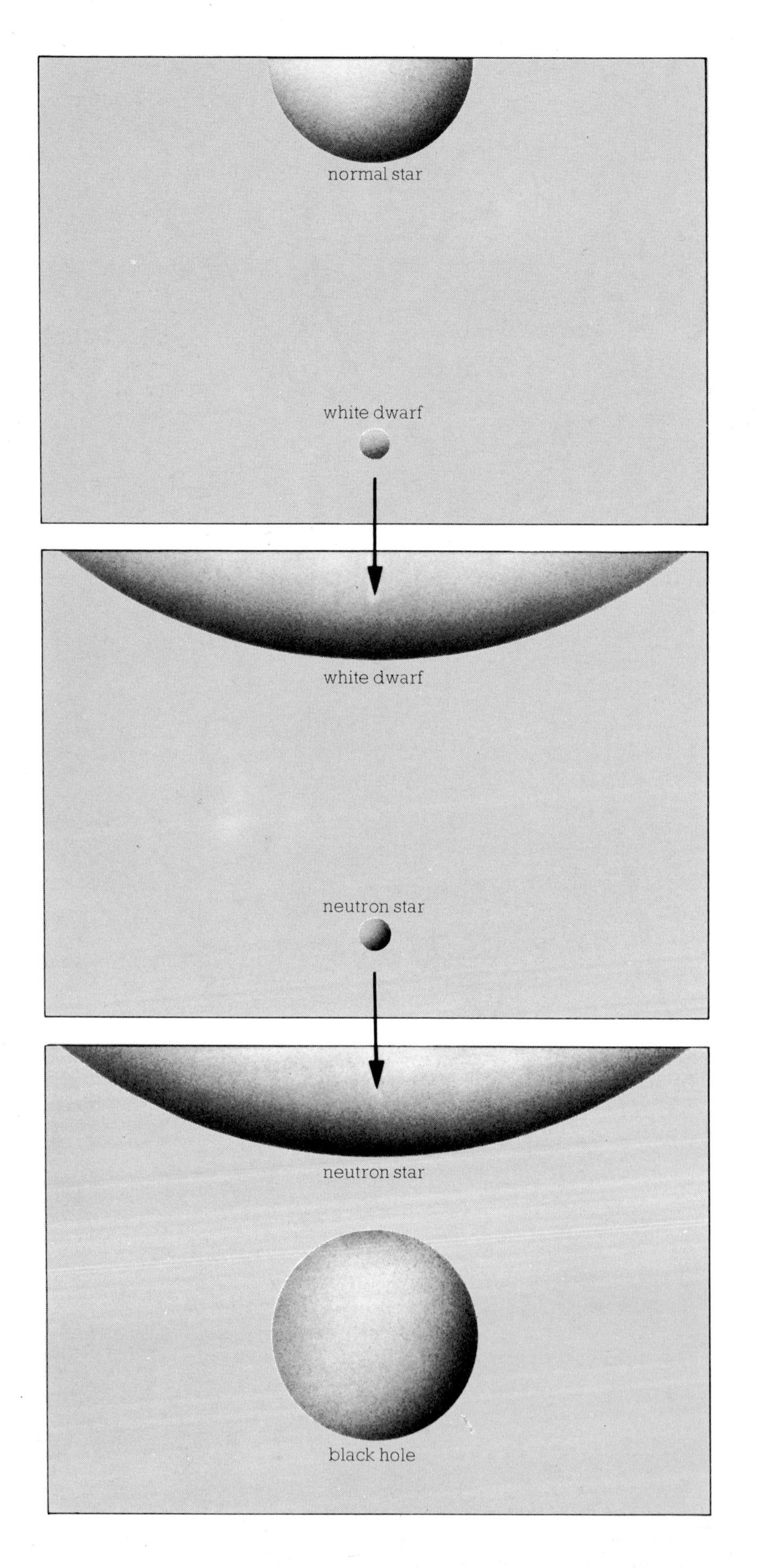


*The sizes of a normal star, a white dwarf star, a neutron star, and a black hole compared.*

*A supernova erupts in a distant spiral galaxy. For a few months a dying massive star shines out with the light of millions of Suns before dwindling to a feeble flicker. In the final long exposure photograph we see the galaxy clearly, but the star that exploded has faded from view.*

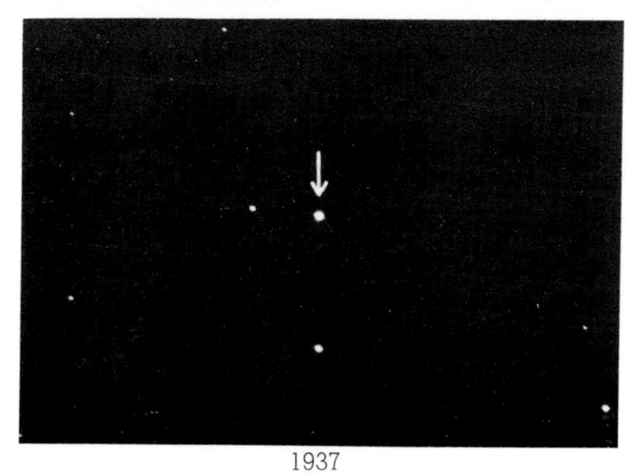

1937

1938

1942

flash of radio waves on each rotation. This type of neutron star is called a pulsar. A neutron star will not form without first causing a stupendous explosion. In its last few seconds a dying star may flare up as a supernova. For a few days it outshines entire galaxies. The central part of the supernova forms a neutron star.

A supernova seen in AD 1054 is now known to have made the famous Crab Nebula. At the heart of the cloud there is a pulsar which spins thirty times per second.

It is not possible for neutron stars to be greater than two Sun masses. A dying star of ten Sun masses, for example, suffers so badly from the gravity it generates that no force is large enough to withstand the collapse. When a star like this has shrunk to a couple of kilometers, its gravity is so huge that the escape velocity is greater than the velocity of light. Anything—rockets, particles, light, or radio signals—trying to escape cannot do so. Gravity is so strong that it drags everything back to the star. As far as we are concerned, the star has become a black hole. We cannot see it because no light can escape from it.

Black holes can be detected in space because their mass continues to attract other matter. So a black hole orbiting another star has a drastic effect. In fact, X-ray telescopes have detected rays from gas that is rushing down black holes. Once the matter has fallen down one of these cosmic whirlpools, it appears to vanish from our Universe, although we can still feel its gravitational pull.

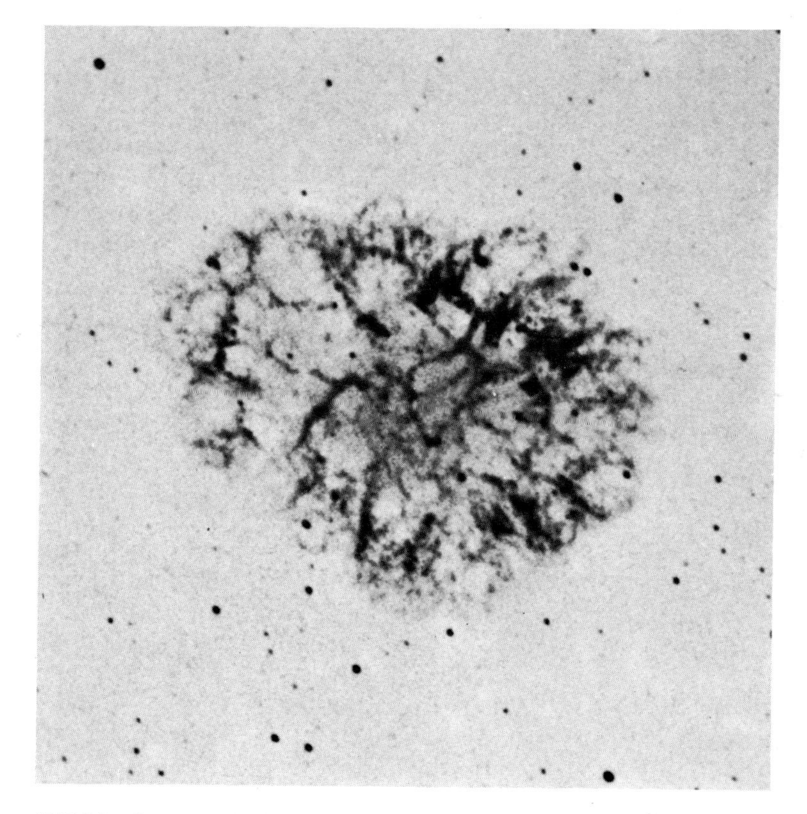

*Within the tangled wreckage of the Crab Nebula a tiny neutron star spins thirty times a second. The Nebula and neutron star are the remains of a star that exploded nine hundred years ago.*

# Gas Clouds between the

Although the stars in the night sky look close together, they are really separated by immense stretches of space. Most of the gaps between the stars look dark, but space is not in fact completely empty. There are atoms of gas and particles of dust floating throughout space. This matter creates a thin haze that dims the light from distant stars. Astronomers call the space dust and gas interstellar matter. It is much thinner than our air. A cup of air contains around a thousand million million million atoms, but a cup of the stuff in deep space would contain only five hundred atoms!

In places the gas and dust have collected together, or been swept up by gravity, to form thicker clouds. Many of these clouds are so thick that they completely shut off the light from stars behind them. On a crystal clear night, when the Milky Way stands out, you can see dusty clouds that make dark patches where they break up the band of silvery light.

Besides the dark dust clouds, there are also glowing clouds of gas that emit pinkish light. These are among the most beautiful objects in the sky, and include the Carina Nebula and the Orion Nebula. Nebula derives from the Latin word for cloud and it has the plural form nebulae.

*The Trifid Nebula is a dense cloud of hydrogen, laced with dark lanes of dust. Very hot stars deep inside the gas make it glow red.*

*In southern skies the most glorious nebula is in Carina. Many bright stars can be seen shining right through this magnificent hydrogen cloud. At the center of the Carina Nebula, there is a very strange star that has greatly changed in brightness over the last century. It was once the second brightest star in the southern sky but can now only be seen with a telescope.*

# Stars

*The Doradus Nebula lies in the Large Magellanic Cloud, a star family just beyond the Milky Way. This nebula is larger than any in the Milky Way.*

In winter, Orion the Hunter is one of the easiest star groups to recognize. Just below the row of three stars that make up his belt there is a fuzzy patch of light. This is the Great Nebula in Orion, one of the few that can be spotted without a telescope. Field glasses or a small telescope will show you that it glows with a soft greenish light.

When gases are made to glow, each different gas gives out light of a different color. Everyone has seen the orange colored neon lights used in advertising signs; in Europe sodium lights are used to illuminate main highways. Neon is a gas that glows orange when an electric current is passed through it. Sodium lights look yellow.

In color photographs the shining nebulae of space usually look pinkish or purplish. Most of the gas in space is hydrogen and the bright stars inside the nebulae cause the hydrogen to glow in a similar way to neon. Hot stars give out invisible ultraviolet rays which make the hydrogen gas shine with the beautiful pink-red light when the rays travel through it. As well as these bright clouds, there are other nebulae that act as cosmic mirrors, reflecting the visible light from stars that are near to them.

An exciting discovery about the gas clouds of space is that new stars are being made inside them all the time. Astronomers have actually seen new stars turn on inside the Orion Nebula. A new star begins when particles of gas and dust collect together into a huge ball. The pull of gravity squeezes the ball tighter and tighter, and at the same time it gets hotter and hotter. Eventually, the ball is hot enough to start the nuclear reactions, and it then continues to shine of its own accord. Then it settles down to a lifetime of being an ordinary star, spending much of its time in an existence similar to our Sun's.

In the Milky Way—the great family of stars to which the Sun belongs—about nine-tenths of the material has already formed into stars. The other one-tenth is the gas and dust spread out between the stars. Within this material new stars are being created. The new stars are generally grouped into open clusters, like the Pleiades.

# Dust Clouds and Life

Dark patches cut through the silvery haze of the Milky Way; they are enormous clouds of space dust and cold gas. Surprisingly, perhaps, space is a dusty place. Between the stars lies a sprinkling of sand, grit and soot. When we look at the stars near to the Sun, the space dust is not noticeable. But stars far away, more than a thousand light years of distance, shine with a reddish hue. You've noticed how our Sun looks very red when it is close to the horizon; that is because its light is passing through murky layers in our atmosphere. Similarly, when starlight has to travel through hundreds of light years of interstellar dust it gets redder in directions where the dust is particularly thick. Stars tens of thousands of light years away can scarcely be seen even through large telescopes. We cannot see deep into the heart of the Milky Way for this reason.

Sometimes the dust collects together with gas and forms a dark nebula, that is a cloud in space which emits no light. The most famous examples of this are The Coalsack, an astonishingly black void in the southern Milky Way, and the superbly-shaped Horsehead Nebula in Orion. In the Horsehead Nebula, the dramatic horsehead is silhouetted against a flood of light reflected by another nebula.

Within the Orion Nebula there is a large dust cloud that emits radio waves. Similar dust clouds occur in the constellation Sagittarius, towards the central section of the Milky Way. Radio astronomers have discovered that many of the signals from these dust clouds are caused by molecules. The detection of certain radio signals from the clouds tells us which molecules are in the clouds.

Over thirty different types of molecule are present in thick dust clouds. Some are simple substances, for example, carbon monoxide, which is made of carbon and oxygen. Others are more complicated, being made of several atoms; formic acid is an example—on Earth it is the substance that puts the sting in nettles and bees!

All life on the Earth is based on the structure of one huge molecule, named DNA. The molecule carries in code form instructions on how to duplicate itself. The molecular assembly line uses simpler components, such as sugar and protein, to build up DNA. These substances can be manufactured from even simpler molecules. In the dust clouds of space the simplest molecules needed to start the assembly of the complex substances exist in great profusion. Perhaps the first living things on Earth grew from these elementary molecules. It is within the densest clouds that new stars and planets are born. When the Earth first condensed it may have been enriched with life molecules from a cosmic dust cloud.

*The Horsehead Nebula in the constellation of Orion is a dense cloud of dark gas and dust. Behind the nebula brilliant young stars emit a flood of light that makes the horse's head stand out dramatically.*

*The beautiful Orion Nebula is a birthplace of the stars. Hot hydrogen gas flushes a glorious pink. Dark clouds thread this nebula. In the lower right corner, an immense cloud of dust is a haven for molecules. In 1978 an entirely new star began to switch on in Orion, thus proving that it really is a cosmic nursery.*

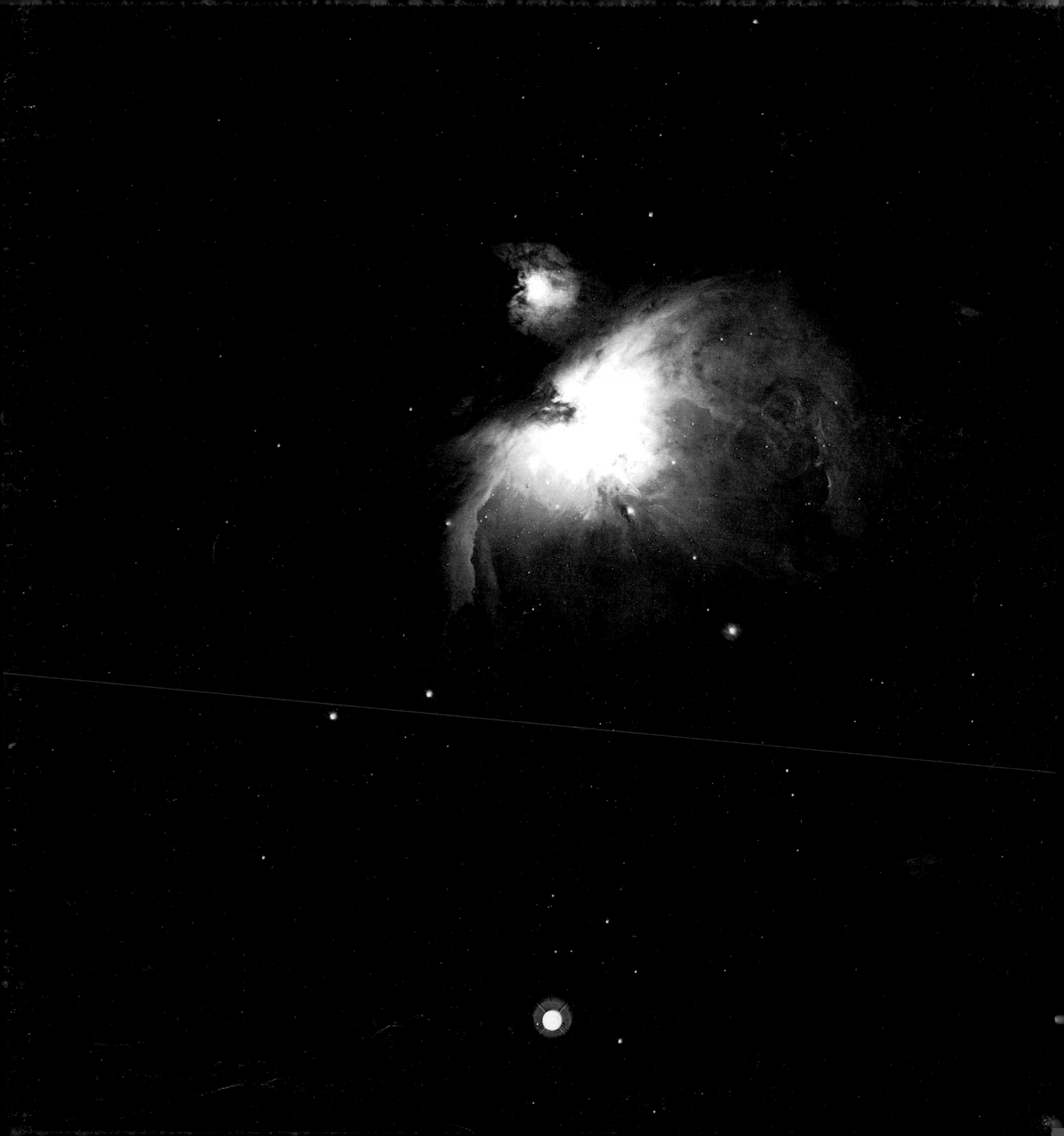

# The Milky Way

On a clear dark night stars spangle the sky, looking near enough to touch. In fact, most of the stars visible to the naked eye are within a thousand light years. Apart from the twinkling stars, a faint band of light called the Milky Way stretches across the heavens. This silvery haze of light is tens of thousands of light years away. With field glasses, or a small telescope, the Milky Way is revealed as a dense crowd of countless thousands of faint stars. Within the band of the Milky Way, the combined light of millions of remote stars is just sufficient to register in our eyes.

Away from the direction of the Milky Way there are far fewer faint stars so that no misty glow is detectable.

Since the Milky Way forms a complete circle round the heavens, parts of it are visible from everywhere on Earth. Among the main constellations through which it passes are Cassiopeia, Perseus, Auriga, Monoceros, Vela, Crux, Scorpius, Sagittarius, Aquarius, and Cygnus. The richest star fields lie in the southern Milky Way, forming a splendid sight in Australasia and southern Africa. For northern observers, the Milky Way is at its finest late on summer evenings when Cygnus, the Swan, is overhead.

But what precisely is the Milky Way? In fact, it is our view, from the inside, of the great starry Galaxy that we live in. Our Galaxy has perhaps one hundred billion stars altogether, and because we live inside it is not easy to visualize the shape directly. In fact, the Milky Way Galaxy is a gigantic pinwheel with two starry arms that wind round the central part several times. The distance across our Galaxy is 100,000 light years. From Earth it takes 30,000 light years for a radio message to travel to the center of the Galaxy. And to count all the stars, one by one, would take a thousand years at the rate of three each second.

The brightest part of the Milky Way is in Sagittarius, and from this region radio and infrared telescopes detect strong signals. At the chaotic center of the Galaxy there may be a huge black hole, somewhere in the direction of Sagittarius, that is busily gobbling up stars and planets that drift in too close!

Our Galaxy is rotating round its central regions, but it does not turn like a solid wheel. Stars near the centre make one orbit in only a few million years, and yet out near our Sun a single circuit takes 250 million years. Since its formation, the Sun has only managed twenty round trips, and since people have appeared on Earth the Sun has made less than one-hundredth of a galactic orbit.

The slow wheeling of our Galaxy, with the inner sections continually overtaking and lapping the outskirts, means that the stars themselves are steadily drifting about the sky. Within a few thousand years the constellations will look different as a result of this restless starry movement.

*Our Galaxy is similar to a vast pancake or pinwheel and the Sun is located on one edge, thirty thousand light years from the center. Open star clusters are found in the main disk. In a halo surrounding the Galaxy we find a hundred or so globular clusters. These star families were formed ten thousand million years ago. They got stranded beyond the present boundary of the Galaxy when a ball of matter that became the Milky Way Galaxy flattened itself into a disk.*

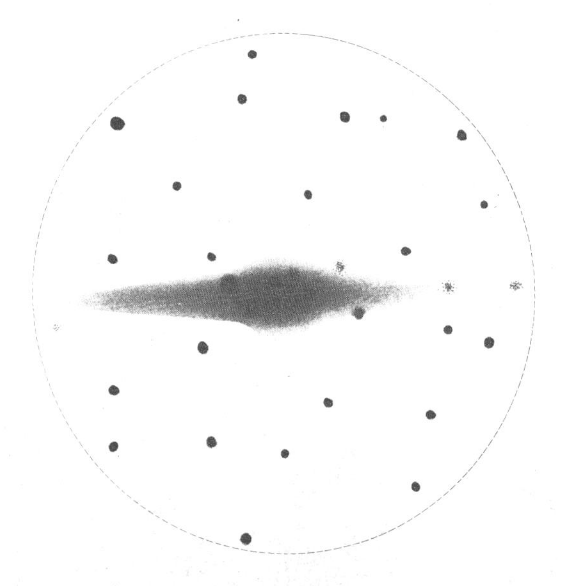

*The Milky Way spans the heavens in this wide-angle photograph of the northern sky. All round the edge of the photograph are blurred images of trees and lights on the horizon. The hazy light of the Milky Way is due to the combined effect of a host of remote stars. Our Galaxy, which we see as the Milky Way, is a vast family of one hundred billion stars held loosely together by the force of gravity.*

# The Galaxies

Our own Galaxy is huge, but beyond it stretches a whole Universe that is unimaginably vast. Throughout this space there are scattered countless millions of other galaxies, some like our Milky Way system, and others completely different. Just as the Sun is only one among the myriads of ordinary stars, so our Galaxy is just one ordinary member of the universal family of millions of galaxies.

Astronomers did not even realize that there were galaxies outside the Milky Way until two centuries ago, when William Herschel made his famous survey of the sky. His telescopes kept bringing misty nebulae into view. Some of these, Herschel reasoned, must be Island Universes, far outside our own system of stars. In the early twentieth century careful studies of variable stars confirmed Herschel's ideas. Even the nearest galaxies are millions of light years distant from us.

Galaxies come in all shapes and sizes. Mostly they are either spiral, like the Milky Way Galaxy, or elliptical, something like a lemon. A small proportion is irregular in shape; these misfits seem to have been disturbed in some way. The number of stars within a galaxy may vary from a few hundred million up to perhaps a million million in a giant elliptical galaxy. The Milky Way is a large galaxy both as regards size and number of stars.

At one time it was thought that elliptical galaxies gradually became flatter and then sprouted spiral arms. This is completely wrong: there are two main types of galaxy that are made in different ways and neither kind can turn into something else!

*Galaxies come in many shapes and sizes. All of them are thousands of light years in diameter and they contain millions of stars.*

Apart from visible stars, the spiral galaxies also contain gas and dust. When a spiral is turned edgewise towards us, as some of them are, dense clouds of dust can be seen in the spiral arms. Elliptical galaxies don't have much gas: it has either all turned into stars by now or was blown away when the galaxies first formed twelve thousand million years ago.

The two nearest galaxies are the Large and Small Magellanic Clouds, named in honor of the great naviga-

*The famous Andromeda Nebula is a magnificent spiral galaxy two million light years from our Milky Way. Notice that it has two smaller elliptical galaxies in orbit round it.*

tor Ferdinand Magellan, who noticed them during his round-the-world voyage. They can be seen from the southern hemisphere as large hazes of light. Both of these irregular clouds of stars are dwarf galaxies 175,000 light years from us.

The great galaxy in the constellation of Andromeda is similar to the Milky Way Galaxy. It is about two million light years out in space and is the most remote object that you can see with your own eyes. Imagine, when you gaze at Andromeda you're seeing light that started its journey to you before people walked the Earth! Yet our telescopes can now photograph galaxies that are three thousand million light years distant. Suppose we tried to send a radio message to creatures in these far-off galaxies. It would take six thousand million years for the message to arrive and for us to receive the reply. Clearly we shall never be able to converse with other galaxies. All we can do is look, listen, and wonder.

*The galaxy M83 in the constellation of Hydra is a spiral which faces us wide open. Brilliant young stars and glowing gas clouds stand out like gleaming jewels in the spiral arms.*

# Clusters of Galaxies

The spiral galaxies of Andromeda and Triangulum are two members of a local family of galaxies. Other members are the Milky Way and the Magellanic Clouds. About fifty close neighbors of the Milky Way are visible through telescopes. This local family of galaxies lives in a sector of the Universe that measures about ten million light years from end to end. Astronomers call a crowd of galaxies grouped together in this way a cluster of galaxies. These clusters of galaxies are the largest single entities in the Universe.

Within our own local group of galaxies the members are loosely attached by the force of gravity. To venture beyond our corner of the Universe, an immense empty gulf must be crossed before the next great cluster is reached at a distance of fifty million light years. And beyond this, Schmidt telescopes have photographed more and more clusters of galaxies, growing ever fainter with increasing distance.

If you wished to make a rough map of a strange country, you could do it by marking down the positions and sizes of towns and villages. This would give the general picture a lot faster than if you laboriously mapped out the location of every house and office block. Similarly, astronomers in working out the map of our great Universe find it's sensible to concentrate on the huge starry cities—the giant galaxies and huge clusters. By measuring their sizes and distances, astronomers hope to find out what kind of Universe we inhabit. A sure answer to this research is not yet known.

An intriguing problem of the clusters of galaxies is this: how much mass do they contain? By seeing how the galaxies are moving inside a cluster, scientists can deduce how much material the cluster must contain. Usually this comes out to a lot more matter than is visible as ordinary galaxies. However, X-ray telescopes have found that clusters contain an exceedingly hot gas that is invisible in optical telescopes. This X-ray gas may account for the "missing" material in great clusters. The density of this gas is very tiny—less than one atom per cubic meter—and yet a cluster is so enormous that all the gas atoms put together weigh more than the galaxies!

*A few dozen members of a galaxy family in the constellation Centaurus.*

*The great cluster of galaxies in the constellation of Virgo.*

*An impression of what our own family of galaxies might look like if viewed through a telescope located in the next nearest cluster. Three spiral galaxies, the Milky Way Galaxy, Andromeda and Triangulum, dominate the scene. A host of smaller galaxies keeps these giants company.*

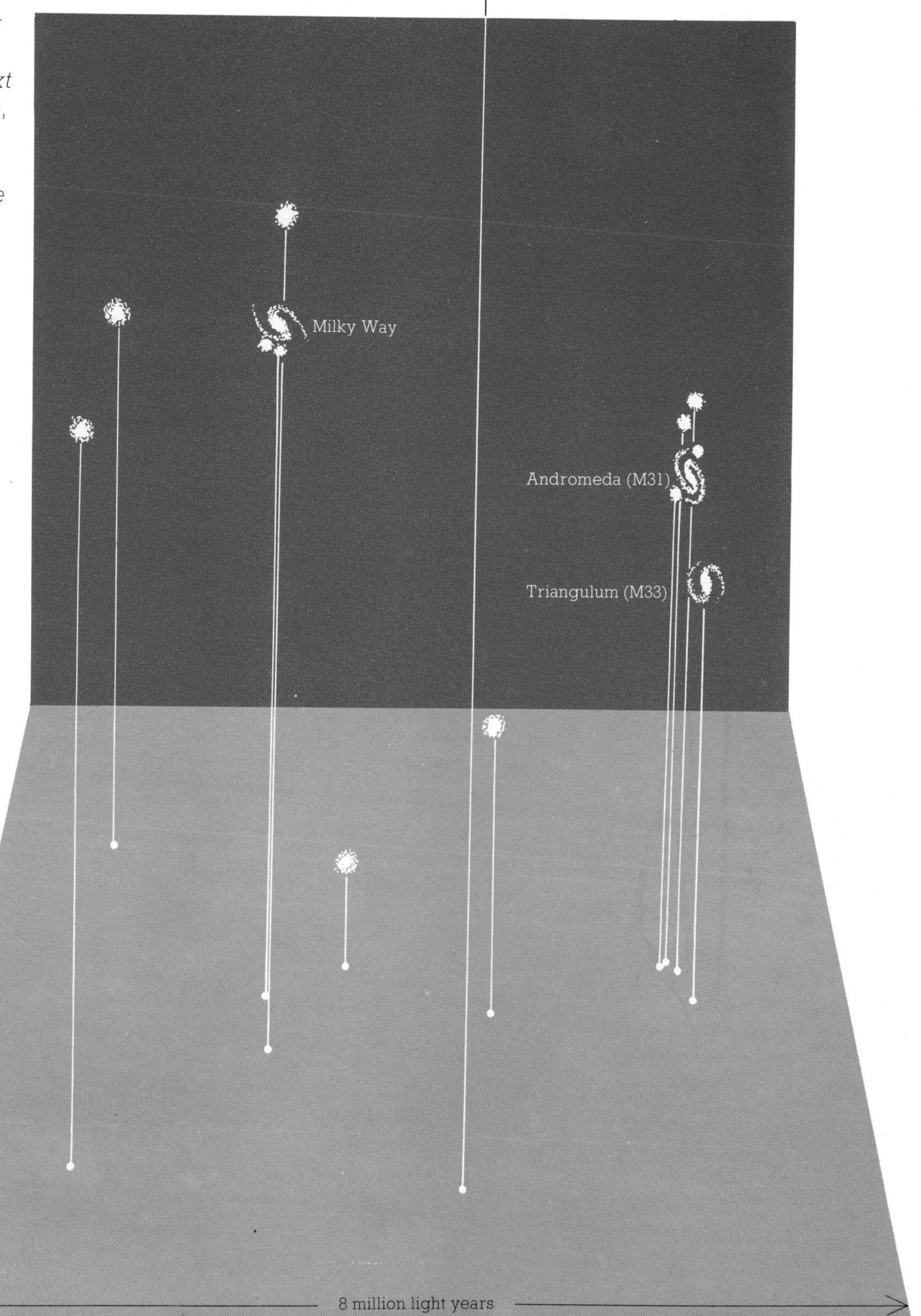

# Exploding Galaxies

Far out in the Universe, entire galaxies are exploding. As the center of a galaxy tears itself to shreds, stars are destroyed in a flash. In certain galaxies, the brilliant light from the explosion is all we see. The starlight is insignificant compared to the grandeur of a galactic catastrophe.

Radio astronomers made the first discoveries of exploding galaxies over thirty years ago. They found a strong emitter of radio waves, as bright as the radio waves from our Sun, in the constellation of Centaurus. Detective work showed that the radio waves were coming from a huge galaxy twelve million light years away. On each side of this galaxy there are two clouds of electrically charged particles. These particles crash through a magnetic field and emit strong radio signals as they do so. Each radio cloud is far bigger than our Galaxy. The highly energetic clouds were probably thrown out of the Centaurus galaxy in an explosion.

Radio-emitting galaxies shine out like beacons in the remote parts of the Universe. In fact, radio telescopes detect the strongest radio galaxies at distances that are well beyond the reach of present optical telescopes. Studies of the visible light sent out by the nearer radio galaxies confirm the idea that they have turbulent central regions, rocked by explosions.

What kind of energy is powering these astonishing radio galaxies? After thirty years of thought we have an outline of what's happening. Somehow an immense black hole, millions of times bigger than the Sun, is formed. Matter, that is to say stars, planets, and gas, gets sucked into the hole and is partly changed into energy. The energy released is great enough to make the central region of the unfortunate galaxy burst to pieces, sending a flood of electric particles far beyond the galaxy.

Apart from the radio galaxies, astronomers have observed other varieties of turmoil between the galaxies. Just as the Earth and Moon are linked by the force of gravity, so the galaxies sometimes get locked together in pairs. Then tidal forces rip their spiral arms to shreds. Long streamers of stars and gas may get stranded in empty space, far from their parent galaxies. Most of the distorted galaxies are in small clusters, where collisions and near misses are quite common.

*A small cluster of galaxies locked together by gravity. Large forces have distorted the galaxy shapes.*

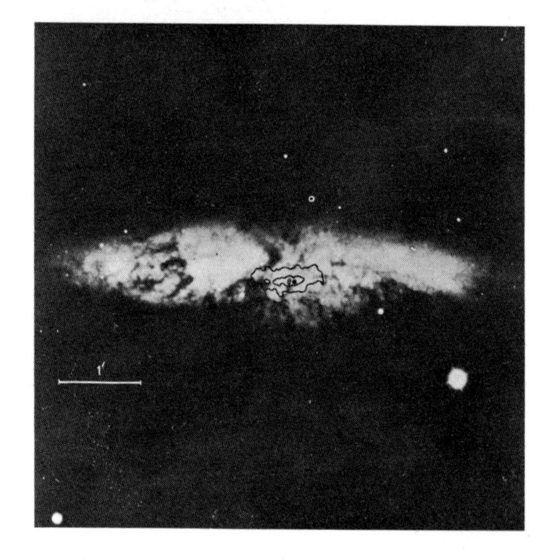

*Galaxy M82.*

In the constellation of Ursa Major, there is a very strange galaxy named M82. This object consists almost entirely of billowing clouds of hydrogen gas that have swamped the stars in a flood of pink light. Galaxy M82 may have crashed into an invisible pile of dust lurking in space, or its centre may have exploded a few million years ago.

*Centaurus A is a beautiful radio galaxy in southern skies. It is circled by a band of thick dust. Outside this galaxy there are two stupendous and invisible clouds that act as strong transmitters of radio waves. Perhaps Centaurus A is powered by a black hole millions of times more massive than our Sun.*

*Galaxy Perseus A is a tangle of hydrogen shreds cast aside by an exploding center. This galaxy is still being stirred by violent events. Radio and optical astronomers have seen its central regions change in a few weeks.*

# The Quasar Mystery

When observers followed up the discovery of powerful radio galaxies they stumbled across a mystery. Some of the sources of radio waves could not be matched to visible galaxies. Instead, the optical telescopes detected a tiny point of light in the locations of some radio sources. Although these objects looked like stars at first glance, spectra showed that these "radio stars" were far stranger than any star within our Galaxy. The spectra of the star-like radio sources indicated that they are located at exceedingly great distances, thousands of millions of light years from Earth.

These intriguing and distant objects became known as quasars, an abbreviation of their original name "quasi-stellar radio source". The main properties of a powerful quasar are these: it sends out a great flood of light, as much as a hundred giant galaxies; this energy flows from a region of the quasar that is much smaller than a galaxy, which is why a quasar looks as tiny as a star on a photograph; the spectrum shows that a quasar is thousands of millions of light years away. Many quasars, but not all, send us radio waves; a few emit infrared, ultraviolet and X-rays as well.

Quasars are the furthest objects ever seen. The light from some of them was halfway here before the Sun and the Earth came into existence. In fact, the remotest quasars lie about nine-tenths of the way to the limits of the observable Universe. The light from these far-off objects has travelled to us almost from the beginning of time.

As far as we can tell, the powerhouse in a quasar is not very much larger than our solar system. Certainly the distance from here to the nearest star would be big enough to hold the power station in many quasars. Yet within this small volume, the quasar pours out as much light as a million million stars. How can it do so?

The quasar mystery is to find the source of the energy. Probably a gigantic black hole lives in the center of the quasar. Matter falling into the hole releases the energy that we can see. Only by this means can enough energy be made. Exploding stars and nuclear holocausts are not nearly powerful enough to keep a quasar burning. Only gravity, the force that ultimately controls the Universe, can unlock the vast energy supply that a quasar needs.

Many of the quasars were formed early on in the past history of the Universe. We see them today not as they are "now" but as they appeared long long ago. By now they will have burned out and possibly transformed themselves into normal galaxies. If this idea is indeed true, perhaps some of the galaxies near to us, or even the familiar Milky Way, are in fact dead quasars.

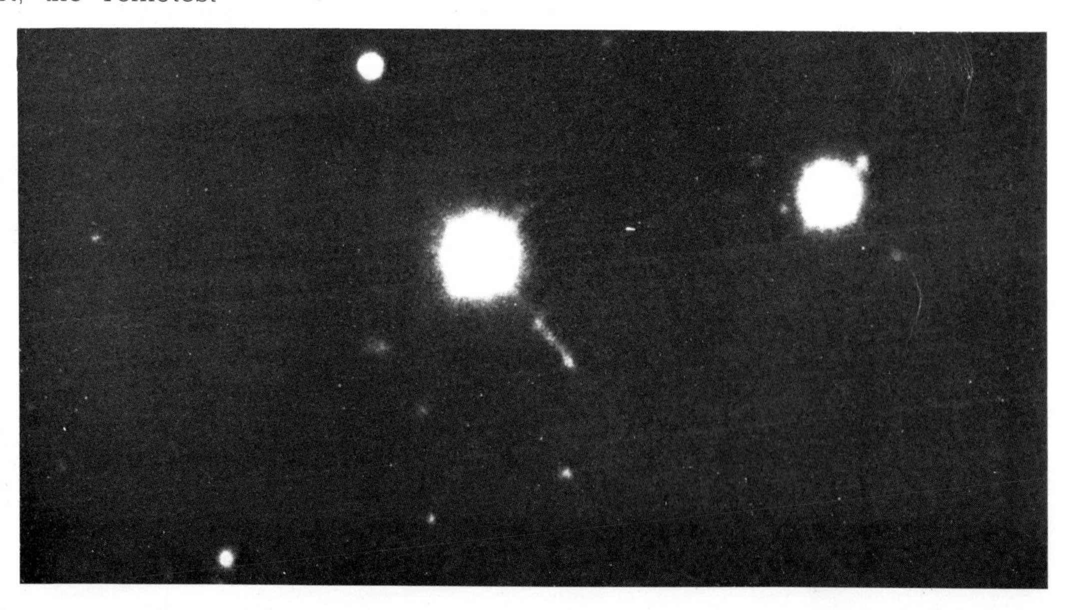

*Quasar 3C273 has blasted out this jet of electrons from its central regions. The electron jet emits an eerie blue light. Within other quasars, jets of matter have been expelled in a similar way. The matter speeding away from a quasar may have a speed as high as half the speed of light relative to the quasar.*

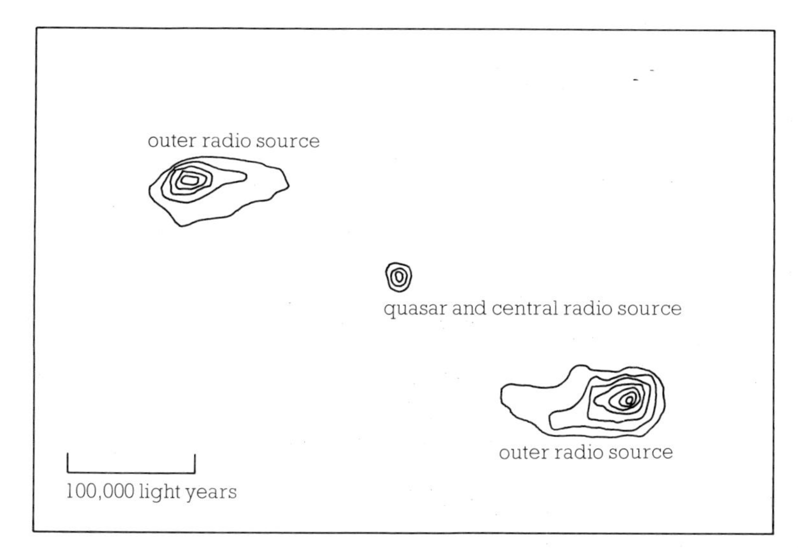


*The radio waves from a quasar occupy a huge volume of space that surrounds a tiny point of light. Quasars are the most powerful radiators of energy known to science.*

*Eventually a quasar must run out of energy, after its central black hole has eaten all the stars within easy reach! Perhaps then it settles down to become an ordinary galaxy.*

*Here is a quasar that apparently lies close to the lemon-shaped galaxy. Notice that the quasar looks like a star (all the other images in this picture are stars) rather than a galaxy. Measurements show, however, that the quasar is very much further away from us than the galaxy is. This quasar sends out as much light as dozens of galaxies. It only looks dimmer on the photograph because it is much further away from us than the galaxy.*

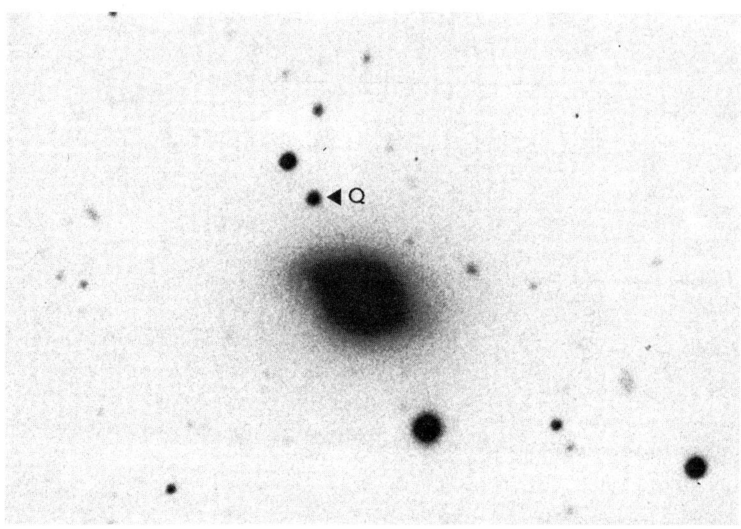


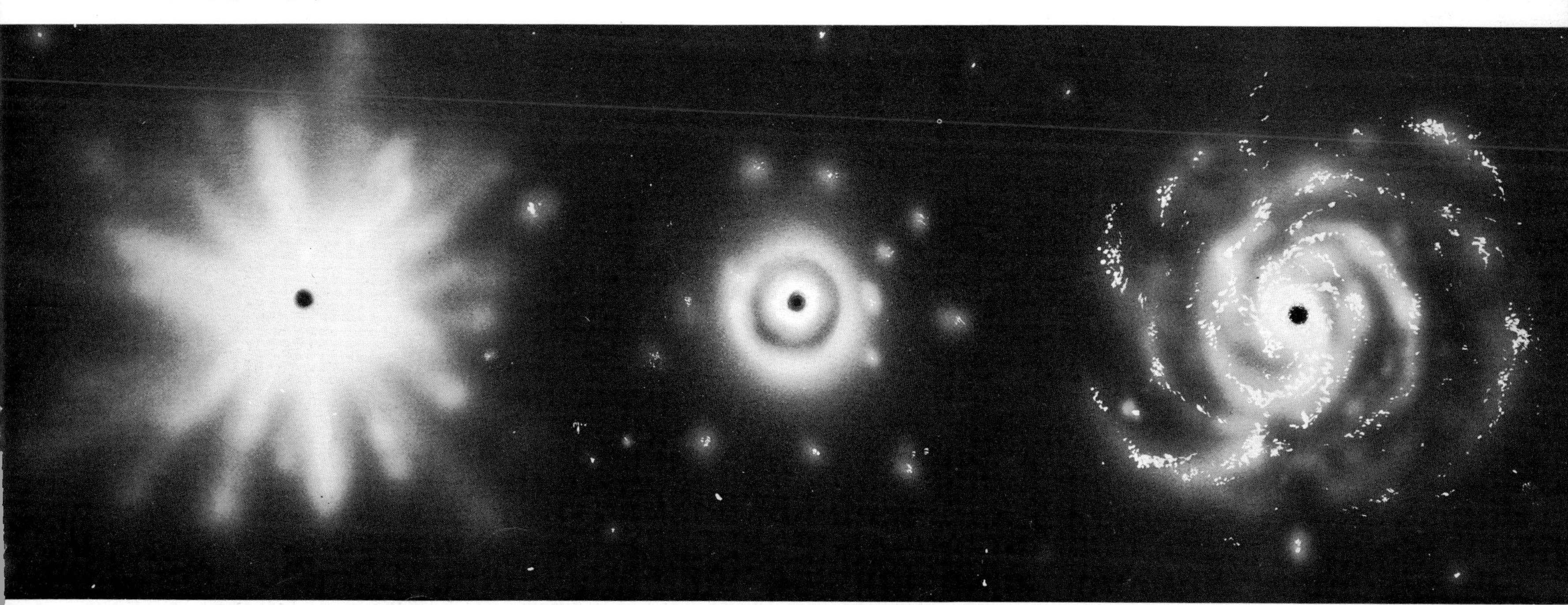

# The Origin of the Universe

We live at the center of unbelievable immensity. In every direction teeming galaxies crowd the telescopes and radio signals flood in from the remotest quasars. We can only wonder how and when and where this all began.

Cosmology is the branch of astronomy which tells us about the properties of the Universe as a whole. Until a few years ago, this was a task that only brilliant mathematicians could solve, using formulae and equations to deduce the nature of the Universe. Now the situation has changed. Many observations tell us about the Universe and its past history.

An important discovery relates the distance of a galaxy from us to its observed speed. All the faraway objects are apparently racing away from our Galaxy. The furthest ones are moving fastest. At a distance of one thousand million light years the galaxies are rushing out at one-tenth the speed of light. Some of the quasars are galloping off at three-quarters the velocity of light. This curious feature, that objects a long way off are travelling out at high speeds, shows that the Universe is expanding. If we back track the paths of the speeding galaxies, we find that they were all in one place thirteen thousand million years ago.

We are not at the center of the expansion, however. No matter where you were in the Universe, you would see everything else trying to get away from your galaxy as fast as possible.

The early phase of the Universe, or the Big Bang that flung out the galaxies, was unimaginably hot, so hot that even today the entire Universe is glowing with the after effects. Radio astronomers have found that space is filled with feeble radio waves. These are literally the dying radio echo of the Big Bang. The radio waves show us that space—empty space—is not totally cold. Rather, the heat from the Big Bang has given the whole Universe a temperature that is 3°C above the coldest possible temperature, absolute zero.

Careful measurements of the speeds of the galaxies show that the Universe as we know it has existed for thirteen thousand million years. Astronomers cannot say what happened before then. Perhaps our Universe emerged from the collapse of an earlier Universe, but we can *never* know for *certain* if this is the case. Our Earth, Moon and Sun have existed for little more than one third of the time since the origin of the Universe. The birth of civilization, the foundation of towns and the organization of human society have taken up less than one millionth of the age of the Universe. What will mankind accomplish in the next millionth? Can the nearest stars be colonized?

In ten thousand years' time there might be human colonies round local stars. But it is hard to imagine how the Universe itself can be colonized. Telescopes have already detected quasars which are ten thousand million light years away. Billions of galaxies sprawl through the Universe, most of them billions of light years from us. The sheer size of our Cosmos is a great barrier to communication and travel. We can only learn about the Universe at large distances and early times by studying the wheeling galaxies and flaring quasars. That is part of the ageless fascination of astronomy, the oldest of the sciences, now passing through one of the most exciting eras in its long history.

*In the initial burst of the Big Bang (top left) no galaxies existed, only fragments of atoms and intense heat and light. After a million years the first threads of hydrogen gas form from the atomic particles (top right). This gas then clumped into cloudy masses greater than entire clusters of galaxies (bottom left). Finally, the galaxies themselves condensed in the clumps about one thousand million years ago (bottom right). Today the galaxies are far apart and the Universe is thinly populated with stars and galaxies. Will the galaxies expand for ever? We do not know for certain.*

# Index





# Acknowledgements

We are very grateful to the following individuals who kindly assisted us by supplying photographs: Dennis di Cicco (Sky and Telescope), Owen Gingerich (Harvard), Keith Tritton (UK Schmidt Telescope), David Malin (Anglo-Australian Telescope), S. Marx (Ondrejov Observatory), Donald Osterbrock (Lick Observatory), James Vette (NASA World Data Center), and Bob Wilkins (JPL, Pasadena).